MATERIALS AND MANUFACTURING TECHNOLOGY

NOVEL APPROACH OF ADDED-VALUE ZINC OXIDE POWDERS FOR POLYMERIC FIBROUS MATRICES WITH ENGINEERED ARCHITECTURES FOR HIGH PERFORMANCE TEXTILES

Materials and Manufacturing Technology

Additional books in this series can be found on Nova's website
under the Series tab.

Additional e-books in this series can be found on Nova's website
under the e-book tab.

MATERIALS AND MANUFACTURING TECHNOLOGY

NOVEL APPROACH OF ADDED-VALUE ZINC OXIDE POWDERS FOR POLYMERIC FIBROUS MATRICES WITH ENGINEERED ARCHITECTURES FOR HIGH PERFORMANCE TEXTILES

NARCISA VRINCEANU
AND
PARASCHIVA POSTOLACHE
EDITORS

nova publishers
New York

Library of Congress Cataloging-in-Publication Data

Novel approach of added-value zinc oxide powders for polymeric fibrous matrices with engineered architectures for high performance textiles / editors, Narcisa Vrinceanu and Paraschiva Postolache ("L.Blaga" University of Sibiu, Romania, and others).
pages cm. -- (Materials and manufacturing technology)
Includes bibliographical references and index.
ISBN 978-1-63483-432-2 (hardcover)
1. Textile fibers, Synthetic--Bonding. 2. Zinc powder. 3. Textiles--Additives. I. Vrinceanu, Narcisa, editor. II. Postolache, Paraschiva, editor.
TS1548.5.N625 2014
677'.02835--dc23

2015026402

Published by Nova Science Publishers, Inc. † New York

CONTENTS

PREFACE

The aim of this monography is to summarize the fundamentals and technical approaches in the processing and behavior of nanostructured materials in the textile industry, to provide a comprehensive assessment of exciting opportunities for diversification in the textile industry, upgrading the existing functions and performance of textile materials, developing smart and intelligent textiles with unprecedented functions.

The book mainly focuses on zinc oxide architectures with regard to their syntheses, characteristics and responses to environmental factors (UV radiation, water and water vapors, and heat, under controlled laboratory conditions). The interdisciplinary attribute articulates it to the design of added value products, as engineered materials with enhanced performance under environmental conditions, comfort and time stability of the garments.

Several new concepts and models are introduced by synthesizing ZnO nano and bulk (micron) powders using simple and facile methods. Two new ZnO nanostructures are discovered and analysed.

In addition, the monography was oriented in achieving a textile nanocomposite with a surface architecture allowing blocking/fixing/adhesion of different guest compounds, resulting in different properties like: water repellence/washing resistance, antimicrobial activity, photoprotection, etc.). Consequently, the particles' size plays a primary role in determining their adhesion to the fibers.

The book promotes a comparison/antagonism between size-controlled ZnO architectures – nano and micron powdered, capped in two different polymeric matrices, but conferring the same type of comfort performance to garments.

The experimental methodologies mainly involved in this work are:

- *in-situ* synthesis of ZnO nanoparticles on linen fibrous supports (yarns) having a certain concentration of mono-chloro-triazynil-β-cyclodextrin by using the hydrothermal method with the assistance of two surfactants;
- padding of polymeric fibrous matrices with ZnO nanoparticles emulsion prepared by means of a clay named montmorillonite;
- synthesis of ZnO nanoparticles by wet chemical method (sol-gel method) using zinc nitrate and sodium hydroxide as precursors and soluble MCT-β-CD, as a stabilizing agent;
- preparation of an inorganic–organic material, based on ZnO microparticles impregnation of polyester textiles.

Various techniques for the measurements of the composites were used: X-ray Diffraction (XRD), Scanning electron microscopy (SEM), and FT-IR spectroscopy. To check the presence of various elements EDX spectroscopy was used as well.

The particles size measurement was done by XRD Scherer's formula. SEM micrographs revealed the morphology of the specimens. EDX confirmed the presence of both Zn and O.

A novel model of entrapment of ZnO particles into MCT-β-CD cavities, in order to obtain structures that might be applied onto polymeric fibrous matrices providing hydrophobic features, was obtained.

Moreover, ZnO-fabric composites with tuned morphological and adsorptive properties, high homogeneity of particles distribution and high surface activity were obtained. By ZnO impregnation of polyester matrices, an inorganic-organic hybrid polymer material, with enhanced UV-protection properties and high hydrophobicity was created.

The novel idea the book initiates is that similar comfort features could be attained, by using two different ZnO sized architectures – nano and micron/bulk, but inserted in two distinct polymeric matrices:

- Increased hydrophilicity/wetting attributes of hybrid composites consisting of zinc oxide nanoparticles introduced with the assistance of two different surfactants in the polymeric matrix (MCT-β-CD grafted linen yarns).
- Enhanced thermal stability of hybrid composites made of nano-ZnO particles applied on cellulosic fibrous matrices (linen supports/fabrics), with the assistance of MMT and MCT-β-CD.

The original and relevant side of the studies with regard to the interdisciplinary view offers an important contribution to stable superhydrophobic ZnO modified textile surfaces manufacturing in a simple and scalable manner allowing, surface tailoring and water barrier attributes with direct implication in comfort performance. Besides, the hybrid item based on ZnO microparticles impregnation in a polymeric matrix is facile and easier to implement in practice. Moreover, the composite is a compromise between comfort and ZnO treatment stability for improved antimicrobial properties.

INTRODUCTION

Narcisa Vrinceanu[1,*], PhD, Eveline Popovici[2], PhD and Mirela Petruta Suchea[2,3], PhD

[1]"Lucian Blaga" University of Sibiu, Romania

[2]"Al.I.Cuza" University of Iasi, Romania

[3]"Al.I.Cuza" University of Iasi,

Center of Materials Technology and Laser, School of Applied Technology, Electrical Engineering Department, Technological Educational Institute of Crete, Heraklion, Greece

NANOTECHNOLOGY OVERVIEW AND DEFINITION

Over the past decades nanoscale science and technology have emerged as the forefront of science and technologies and continues to draw worldwide attention because it is widely perceived as offering huge potential in a wide range of end uses.

Richard Feynman, Nobel Prize winning physicist, is recognized as one who envisioned in his well-known lecture entitled "*There is plenty of room at the bottom.*" (1959) the numerous opportunities associated with the manipulation and control of particles at small level.

Since then, a number of key scientists predicted the birth of what is called today Nanotechnology.

Remarkable contributions in NANO-evolution in materials engineering are presented in the following table Table 1.1.

* Corresponding author address. Email: vrinceanu.narcisai@ulbsibiu.ro.

Table 1.1. NANO-evolution in materials engineering

Nano-engineering contribution	Definition/effect	Functional value-added textiles	References
Norio Taniguchi (Tokyo University of Science)[1]	processing of separation, consolidation and deformation of material by a single atom or a single molecule.		[173]
	controlling atoms and molecules at nanometer scale	functional materials, devices and systems	[32]
	synergy between nanotechnology and the textile industry, exploiting the intrinsic properties of nanostructured materials	UV-blocking, antimicrobial, antistatic and self-cleaning properties, water repellence, electro-magnetic protection and thermoregulatory fabrics	[147]
Nano-Tex	started to invest in the development of nanotechnologies modifying the surface of textiles by means of nanostructured materials	fabrication of nanofibers and nanoparticle composite fibers[2]; upgrading existing functions and performance of textile materials and	[200]
real commercial potential for the textile industry	nanostructured materials have a large surface area-to-volume ratio and high surface energy	high durability for fabrics	
using nanosized materials and growth of nanostructures during manufacturing and finishing processes		water repellence; antiodour; soiling-resistance; antistatic, anti-infrared and flame retardant properties, dyeability, color fastness, as well as durable, or permanent press finished garments; textiles with self-cleaning properties, UV-blocking, deodorizing, thermal regulator, memorizing shape	[70]
two ways in synthesis of nanostructured materials: top-down approach and bottom up approach[3]	classification schemes for nanostructured materials[4]	nanostructured zero-dimensional (0D)[5], one-dimensional (1D)[6], two-dimensional (2D)[7] and three-dimensional (3D)[8] nanomaterials	[129, 24, 130, 131]

[1] 1974.

[2] developing intelligent textiles with completely new characteristics and functions.

[3] Structural elements in 0D, 1D and 2D nanomaterials can be distributed in a liquid or solid macroscopic matrix or be applied on a substrate.

[4] The most important criterion of the classification is the dimension of the array of constituents.

[5] Uniform particles arrays (quantum dots), heterogeneous particles arrays, core–shell quantum dots, onions, hollow spheres and nanolenses, nanocluster materials and nanodispersions in which nanoparticles are isolated from each other.

[6] Ideal systems for exploring a large number of novel phenomena at the nanoscale: nanowires, nanorods, nanotubes, nanobelts, nanoribbons, and hierarchical nanomaterials.

[7] Films (coatings) with nanometer thickness.

[8] Powders, fibrous, multilayer and polycrystalline materials in which the 0D, 1D and 2D structural elements are in close contact with each other and form interfaces.

NANO stands for an "umbrella" term covering a wide range of technologies concerned with structures and processes on the nanometer scale, and in the same time is an unavoidable part of our future.

Nanotechnology - the ability to engineer matter at the atomic, molecular, and supramolecular level - is the creation and utilization of materials, devices, and systems through the control of matter on the nanometer-length scale; it is the generation of larger structures with fundamentally new molecular organizations. Starting from the premise that properties can dramatically change when adimension is reduced to the nanometer range, the manipulation of a material at the atomic and molecular scale to create new functionality and properties is a deep scientific, productive and modern concept. Considering the nanometer scale, the surface-to-volume ratios of materials become large and their electronic energy states become discrete, leading to unique electronic, optical, magnetic, and mechanical properties of the nanomaterials.

According to the intensification of global competition, textile manufacturing companies are competing for a significant share of the global market by developing new technologies and new products.

The main objectives of the leading textile industries are value added applications such UV-blocking, antimicrobial, antistatic and self-cleaning properties, water repellence, electro-magnetic protection and thermoregulatory fibrous matrices, and so on. Textile polymeric matrices are some of the best platforms for deploying nanotechnology; the synergy between nanotechnology and the textile industry judiciously exploits the intrinsic properties of nanostructured materials [79].

The first work on nanotechnology in textiles was undertaken by Nano-Tex [148] and later, progresively textile companies started to develop of nanotechnologies modifying the surface of polymeric matrices by means of nanostructured materials.

The nanotechnology's use in textiles can be subdivided as follows:

1. fabrication of nanofibers and nanoparticle composite fibers,
2. surface modification of fibers and fibrous matrices for the purpose of upgrading existing functions and performance of textile materials and
3. developing intelligent textiles with completely new characteristics and functions.

Real commercial potential of nanotechnology for the textile industry is well-known, due to the fact that conventional methods used to induce different properties to fibrous matrices often do not lead to permanent effects, and will lose their functions after laundering or wearing. Nanotechnology alternatives can provide high durability for fibrous matrices, because nanostructured materials have a large surface area-to-volume ratio and high surface energy, thus presenting better affinity for fibrous matrices and leading to an increase in durability of the function [35].

The research in the textile area focuses on using nanosized materials and growth of nanostructures during manufacturing and finishing processes in order to impart variable functions and applications such as: antimicrobial properties, water repellence, antiodour, soiling-resistance, antistatic, anti-infrared and flame retardant properties, dyeability, color fastness, as well as durable, or permanent press finished garments, textiles with self-cleaning properties, UV-blocking, deodorizing, thermal regulator, memorizing shape and other properties [120].

Synthesis of nanostructured materials is performed by two ways in: top-down approach and bottom up approach. Generally, the classification schemes for nanostructured materials are based on different criteria: their chemical composition, the technique for their manufacture, or their dimensionality [129]. For the classification the most relevant criterion is the dimension of the array of constituents which form the nanostructured zero-dimensional (0D), one-dimensional (1D), two-dimensional (2D) and three-dimensional (3D) nanomaterials [129, 24, 130, 131].

Zero-dimensional nanomaterials include uniform particles arrays (quantum dots), heterogeneous particles arrays, core–shell quantum dots, onions, hollow spheres and nanolenses, nanocluster materials and nanodispersions in which nanoparticles are isolated from each other. Ideal systems for exploring a large number of novel phenomena at the nanoscale are one-dimensional nanostructured materials, such as nanowires, nanorods, nanotubes, nanobelts, nanoribbons, and hierarchical nanomaterials.

Two-dimensional nanomaterials are films (coatings) with nanometer thickness. Structural elements in 0D, 1D and 2D nanomaterials can be distributed in a liquid or solid macroscopic matrix or be applied on a substrate. Powders, fibrous, multilayer and polycrystalline materials in which the 0D, 1D and 2D structural elements are in close contact with each other and form interfaces are belonging to the fourth type, meaning three-dimensional nanomaterials.

The main groups of nanostructured materials used in textile industry can be categorized as follows:

- metallic and non-metallic nanostructured materials: silver, copper, gallium, gold nanoparticles, carbon nanotubes;
- oxide nanostructured materials: TiO_2, ZnO, Al_2O_3, SiO_2, nano-layered clay;
- nanocomposites;
- inorganic nanostructured materials loaded organic carriers: cyclodextrin loaded with inorganic materials, nano- and micro-capsules having inorganic nano-particles, metallic dendrimer nano-composites and inorganic nano-particles loaded in liposomes.

Nanoscale Effects

An emerging area of nanoscience and nanotechnology, with broad scientific and commercial perspectives is the application of nanoscale materials and structures. The high surface energy of the inorganic materials, such as metal and metal oxides, mainly determined by size, shape, composition, crystallinity and morphology has attracted lots of attention in functional nano-finishing of textiles. The purpose of the functional nano-finishing is to highlight the added-value technologies based on nanoparticles contribution, focusing on functional textiles with higher quality attributes, such as: antimicrobial properties, water repellence, electro-magnetic protection and thermoregulatory fibrous matrices, antistatic, anti-infrared and flame retardant properties, dyeability, colour fastness, as well as durable, or permanent press finished garments, textiles with self-cleaning properties, UV-blocking, deodorizing, thermal regulator, memorizing shape, and so on.

Table 1.2. Nanoengineered polymeric matrices and their association with high performance added– value textiles

Functional Hygiene	Functional Protection	Durability	Safety	Electronic characteristics	Manufacturing process improvement	Comfort and Design
Antimicrobial	Self-cleaning	Washing durability	Fire retardancy	Supercapacitor properties	Reduction of processing temperature	Softness
Antistatic	UV protection	Dimensions stability	Anti-staining proprieties	Electrical conductive properties	Energy saving	Deodorizing
Air permeability	Stain photo-degradability	Resistance to abrasion	Anti-fungal		Ecofriendly	Anti-pilling
Thermal regulator	Water repellency	Color resistance to light				Color fastness
Hydrophilicity	Stain repellency					Memorizing shape
	Superhydro-philicity					Flexibility
	Superoleo-philicity					High aesthetic appeal
	Dyebility					Easy care

Zinc Oxide Nanomaterials: Current and Future

Zinc oxide (ZnO), as a n-type semiconductor, is a very challenging multifunctional material and has promising applications in solar cells, sensors, displays, gas sensors, varistors, piezoelectric devices, electro-acoustic transducers, photodiodes and UV light emitting devices. ZnO particles can be used as reinforcement phase, wear resistant phase and anti-sliding phase in composites in consequence of their high elastic modulus and strength. Thanks to their current characteristics, ZnO particles exist in anti-electrostatic or conductive phase [4, 109, 181,]. Neverthless, the application of ZnO nanoparticles in coatings system with multi-properties was the topic of few studies. By the traditional coatings technology, i.e., by filling with nanometer-scale materials the nano-coatings can be obtained.

Zinc oxide (ZnO) is a well-recognized coating material able to enhance the functionality of textiles. ZnO powders have a wide range of applications due to the photocatalytic, electrical, optical and antibacterial/antifungal properties. It is also generally regarded as safe material to humans and animals. ZnO was the first material used in form of nanoparticles by incorporation in sunscreen lotions. ZnO has the largest UVA and UVB absorption bands and it is photo-stable [115, 10].

The fact that ZnO nanopowders have an efficient antibacterial and antifungal action was proved. The researches on pathogen agents such as: *Staphylococcus Aureus, Escherichia coli, Streptococcus agalactiae,* etc have demonstrated that ZnO nanoparticles are bio-active [135, 18, 206, 62, 205].

Widely used due to their unique properties zinc oxide (ZnO) powders occurr in a variety of structures, such as photocatalytic, electrical, optical, dermatological and antibacterial [6, 78, 145, 193, 185, 7, 94]. ZnO is known as a UV-blocking material, especially in the UV-A region [34, 10, 93].

As UV-absorbing system ZnO has to be entrapped into different matrices, and to increase the stability of ZnO in different dispersions media and to prevent particles agglomeration various strategies have been developed.

Different methodologies were employed to synthesize hydrophobic ZnO structures among which vapor deposition [69], sol-gel [204], electrodeposition [156] atomic layer deposition [215] or electrochemical anodization [216]. An abundant research concerning various methods of ZnO nanomaterials preparation, such aqueous precipitate synthesis [108], vapor transport, chemical vapor deposition (CVD) [216] was performed.

For the synthesis of zinc oxide nanoparticles water as a solvent and soluble starch as a stabilizer can be used [176, 182]. ZnO nanoparticles were synthesized with the assistance of MCT-β-CD (monochlorotriazinyl–β – cyclodextrin) by using a sol-gel method [169].

Thermal decomposition of zinc salt is one of the versatile ways to obtain ZnO nanomaterials, choosing zinc acetate ($Zn(CH_3COO)_2$) as the precursor due to its high solubility and low decomposition temperature [41]. In order to obtain well-defined ZnO nanomaterials from the precursor of zinc acetate by assisted control, hydrothermal process assisted by microemulsion-mediated has also been reported [189, 39,] as well as sol–gel-derived process [192], etc. All these above mentioned cases have a common aim/objective/purpose – prevention the crystallite agglomeration, control the particle shape, size, size-distribution and crystal phase.

Objectives

This thesis focuses on ZnO controllable architectures and the properties/responses their occurence provides onto the textile polymeric substrates. Emphasis is given on the synthesis and responsive attitude the composites they form.

The uniqueness of this work is that an antithesis between controllable nano-sized and bulk ZnO architectures was devoted, in terms the attributes these structures endowed to the polymeirc fibrous matrices they were applied on.

Subsequently the objectives are divided in terms of scalability of the ZnO powders and the type of fibrous polymeric matrices they were encaved in.

A. Nanopowders onto cellulosic polymeric matrices

Three objectives of the work are as follows:

The first objective refers to the stability of ZnO nanoparticles dispersions, using a polymer matrix, meaning a cyclic oligosaccharide, β-cyclodextrin (*i.e.* MCT- *MonoChloroTriazinyl-β-cyclodextrin*), in the architectural structure of ZnO powders.

The second objective is to investigate novel structure of ZnO-linnen fibrous composites synthesized by means of hydrothermal deposition of ZnO onto linen polymeric matrices, with the assistance of two surfactants, with direct implication onto the relationship between the morphological, structural and chemical attributes and water vapor sorption-desorption behavior.

Thirdly, emphasis is to given on the synthesis and characterization of nanoscaled zinc oxide particles and their application on linen fibrous polymeric matrices, with respect to thermal response/degradation behavior. The thermal treatment of nanocomposites made of nano particle grade zinc oxide and linen fibrous polymeric matrices does not modify significantly their thermal responsivity. The thermal degradation mechanism of the above mentioned structures was also discussed.

B. Micropowders applied onto PES polymeric matrices

This second criterion induces an objective directing the thesis to the evaluation of the surface with regard to wetting characteristics of the composite matrices for improved properties in terms of comfort compared with the unmodified PES. This method can successfully add value to already hydrophobic (enhanced water barrier and UV-protection properties) PES matrix with a direct implication onto comfort features. On the other hand, this surface tailoring by ZnO modification of polyester layers leads to a dramatically decrease in their thermal stability fact which promises improved comfort performance.

The response of this hybrid polymer network to UV irradiation was the photoprotective performance and its time stability, also the hydrophobic character and mechanical properties. Besides, the improvement of antimicrobial properties and quantifying the time stability of the antimicrobial treatment was aimed.

Scope of Study

Covering a wide area of work, a consistent scope of the research is made in order to provide a direction for this thesis.

Finding of practical solution with advantages, as the procedure can be easily standardized, at lower production cost, ensuring a high homogeneity and dispersion of ZnO microparticles into the textile polymeric matrix.

Fabricating of fibrous composites with modified morphological and chemical attributes with high purity, as well as stability in terms of water vapour sorption-desorption. Thus the adhesion between the ZnO nanoparticles and polymer should be strong so that the nanoparticles may not be removed from the host easily due to the formation of an encapsulated complex. The ZnO nanoparticles can this way be attached firmly within the linen fiber surface. This research focuses on new nanostructures of ZnO not discovered or reported by anyone, meaning the lamellar structure similar to biogenic calcites in the presence of echinoderm intracrystalline proteins. The study demonstrates unprecendented the fundamental role playing by polymer involving in the patterning of ZnO composite.

The second scope is the focus on achieving the same results in terms of comfort performance (enhanced water barrier, UV-protection, stable antimicrobial properties), by antithesis between controlable ZnO architectures (nano and micro scaled) and two different fibrous polymeric matrices: natural - cellulosic and synthetic - polyester.

Background of Study

Some progresses have been made in understanding the role of the polymers on nucleation and crystal growth during ZnO synthesis.

Nevertheless, there is still a further challenge to find simple and mild production routes, which will determine the realization of practical applications. For instance, finishing processes by means of ZnO nanoparticles have been among the first commercial application on textiles. But this kind of finishes is not washing resistant, due to poor fixing of these nanoparticles on the textile surface. For this purpose, there is a need of using a polymer matrix with a distinct advantage that is avoiding agglomeration of nanostructures. It is the motivation/rationale of this thesis to uncover the unique characteristics of the nanoparticles of ZnO.

On the other side, washing stability of the complex ZnO-linen polymeric matrices/wetting characteristics is influenced by the type of surfactant used during the hydrothermal synthesis.

Consequently, the missions of this thesis are to discover and characterize new morphologies of ZnO nanoparticles, to identify defect species that dominate the morphology variants, and to study the corresponding shape-dependent impact on the washing resistance.

Besides, it shows regio-specificity and stereo-specificity with respect to the substrate and product. Defects refer to intrinsic types including oxygen vacancies, zinc vacancies, zinc interstitials, oxygen interstitials and antisites species; these intrinsic defects tend to accumulate on the free surfaces (and internal interfaces) of ZnO particles, thereby determine the surface states that are known to be electrically and chemically active.

Insuing, to reveal for the second time the motivation of this research these *shape-dependent* impacts on ZnO potential applications (UV coatings, nanotextiles, photocatalysis and photodetection), the emphasizing the influence of functional polymers *(cyclodextrins)* properties – of forming chemical derivatives and inclusion compounds, namely grafting of a reactive CD derivative and the subsequent inclusion, in the grafted material, of guest compounds resulting new compounds that might have different properties (water repellence/washing resistance, antimicrobial activity, photoprotection protection, etc.).

Cyclodextrins (CDs) present nanocavities (0.57-0.97 nm in diameter) in which molecules of the same size order may be included. As a result, the stability of ZnO nanoparticles dispersions, using such a polymer matrix, in the architectural structure of ZnO powders, was achieved. In presence of both water and heat cyclic oligosaccharide, β-cyclodextrin (*i.e.* MCT- *Mono-Chloro-Triazinyl-β-cyclodextrin*) becomes gelatinized. Gelation of MCT plays a decisive role in controlling morphology of nanoparticles. The prepared nanostructures become pure and water soluble and there are lots of hydroxyl groups attached to nanostructures surface for further functionalization.

The adhesion between the ZnO nanoparticles and polymer through simple wet chemical method is rather poor and the nanoparticles may be removed from the host easily. Because it can be used to obtain products with modified morphological and chemical attributes with high purity and stability in terms of water vapour sorption-desorption, it is assumed that the hydrothermal method can be a more promising way for fabricating nanomaterials [172]. When soluble salt such as zinc acetate ($Zn\ (OAC)_2$) is used, the penetration of Zn^{2+} ions into the interior of polymeric fibrous matrices is facilitated. to The crystallization of ZnO nanoparticles within the polymeric fibrous matrice and to the formation of an encapsulated complex in the hydrothermal environment are possible due to the reaction of Zn^{2+} ions. The formation procedure can be described as follows: first, the coordination compounds are formed through chelation between Zn^{2+} ions and the hydroxyl groups of polymeric matrix fabric. Secondly, under the hydrothermal treatment, the in-situ crystallization of Zn chelate complex occurs and forms a hybrid composite - ZnO- coated polymeric matrix. The ZnO nanoparticles can thus be attached firmly within the matrix surface.

Polymeric fibrous matrices impregnated with ZnO may thus offer enhanced UV protection, and also other advantages. Products made from synthetic yarn have excellent strength properties, dimensional stability and good color fastness to washing, dry-cleaning and light exposure. Nonetheless, there is much interest in designing synthetic fibrous matrices combining these excellent mechanical properties with new features, as combined UV protection and increased hydrophobicity. Such materials are envisaged in many fields of applications, as technical, industrial, medical, military, as well as everyday life uses.

The materials responsivity to water vapors is of interest to a broad spectrum of science and industry. Moisture present in the surroundings interacts with almost all materials. Depending on the material type and the manner it is used, the effects of water can be both harmful and beneficial. Consequently, the thesis aims to determine a correlation between morphological, structural and chemical characterization and the water vapor sorption behavior of the analyzed samples. Consequently, the obtained hybrid composites should find their applicability in textile processing industry subdomains, where a certain level of hydrophilicity/hydrophobicity is mandatory.

The past century has brought a greater sense of comfort, and cleanliness, due to the improvement in qualitative standards of human lifestyles. Even if people are increasingly looking for freshness and a higher level of hygiene in private and public areas, a wide range of micro-organisms coexists in a natural equilibrium with human body and living environments. Moreover, a rapid and uncontrolled multiplication of non-pathogens can seriously compromise the hygiene and wellbeing standards. Due to their broaden applications, the polymeric fibrous matrices have been involved in sustained research efforts for a growing quality of hygienic living conditions. Different combinations of temperature, humidity and other climate factors added to the presence of dust, soil and fat-stains on the polymeric matrices surfaces can transform them in an optimal enriched culture media for rapid multiplication of micro-organisms. This is the case of the occurence of two effects: an uncontrolled spreading from matrice's surface into the surrounding environment with a consequent increase of potential health risks or, at least, of discomfort for the unpleasant odors produced; also degradation phenomena and staining of the textile fibers. In order to avoid these, the application of nanotechnology in the textile industry has increased rapidly. Common polymeric matrices can be coated with multifunctional materials in form of thin films or nanoparticles to get characteristics that increase the protection and comfort of the customer. The manufacture of high value-added products such as smart, medical and protective textiles has also increased rapidly. The approach of deposition of nanoscale sized coatings on polymeric matrices for sensing and monitoring body functions, delivering communication facilities, data transfer, individual environment control and many other applications is intensely studied. For instance, police uniforms made of polyester fibres should present a high level of antimicrobial protection, with respect to the environmental conditions they are worn in. The clothing designed for state police officers should gather, beside a series of criteria set by specific regulations, special requirements regarding antimicrobial protection. But wearing the uniform, the degradation of the antimicrobial treatment is possible. Consequently, this degradation was simulated through repeated washes in the laboratory and its stability was estimated by Zeta Potential measurements. Moreover, the antimicrobial properties as well as its time stability were quantified. The research was conducted on polyester fibrous matrices, previously treated with zinc oxide with various concentrations. The zinc oxide treatment of the materials was realized with respect to ensuring antimicrobial protection

New technologies drawn attention, the so-called "added-value technologies", polarized their efforts on achieving new properties or combinations of properties, e.g. antibacterial, self-cleaning, absorbing UV radiation, deodorizing, thermal regulator, memorizing shape or chameleonic materials, allowing the production of so-called "intelligent textiles".

One of the practical requirements for the safety of garments intended for particular service conditions, as prolonged sun exposure during outdoor activities is the screening against the UV radiation. Transmitted by the fibrous matrix, the UV radiation interacts with the skin and respons by inducing acute and chronic illnesses, including erythema, skin damage and allergies, carcinogenesis and aging. For instance, the UV-A radiation (100 nm - 280 nm) can induce a very high damage of the human skin, the UV–B radiation (280 nm - 320 nm) pigments of epidermal melanoma or skin cancer, and UV-C radiation (320 nm - 400 nm) deep penetration induces skin premature aging, depending on the energy level and exposure duration.

Depending on the material ability to absorb/reflect UV radiation, the UV radiation on the human has severe effects, being in relation to the physical-chemical properties of textile, sun irradiance and wavelength spectrum.

Literature Survey

As clothing supports, polymeric fibrous matrices are designed to protect the body against environmental factors, such as UV radiation, humidity, extreme temperature, bacterial and stain action, etc. In this respect, the textile industry concentrates its efforts to create polymeric fibrous hybrids/fibrous matrices adapted to severe environmental conditions, as to improve the human physiological comfort, by simple and low cost finishing operations [200, 56, 104].

The textile sector requires procedures, starting from the processing of prime materials and ending with the specific finished products. The properties of textile clothing are depending on the processing method. In this regard, allowing preparing of functional materials with controlled properties, in a wide range, from 0.1 nm to micrometers, the nanotechnologies gained much interest in the textiles area. Recently, to obtain various novel fibrous matrices, methodologies like: coating, incorporation or encaving of metal particles (e.g., zinc, gold, silver, platinum, etc.) were tested [188, 195, 33, 74, 153, 89].

Mainly to provide stain-resistance, bactericide properties, or to stop smelly and odorous vapors, the silver nanoparticles have been incorporated into the polymeric matrices of different fibrous matrices, from clothing to bedding or other products [33, 89]. However, in the form of nanoparticles, metals can be harmful to health, if they are absorbed by the human body [84]. In order to avoid such problems, continuous searching for alternative coating/incorporating materials is compulsory.

For instance, in order to improve properties of textiles against UV radiation as well as for antimicrobial protection of polyester and cellulosic matrices, nanoparticles of natural zeolites have been used, and nanoparticles of metal oxides, in particular titanium dioxide and zinc oxide, for UV-blocking or self-cleaning properties [9, 152, 80, 106, 1, 97]. Zinc and titanium oxides are nontoxic and chemically stable at high temperatures exposure.

Research has been oriented to the effect of synthesis of zinc oxide onto cellulosic fibrous matrices previously grafted with MCT-β-CD (MonoChloroTriazinyl–β -CycloDextrin) via a hydrothermal process, onto both wetting attributes and hydrophilicity [169].

By filling with nano-materials, both structure and functional attributes of coatings can be modified: super-hardness, wear resistant, heat resistance, corrosion resistance, and about function, anti-electrostatic, antibacterial, anti-UV and infrared radiation [44, 107, 161, 60].

Powerful emerging multidisciplinary approaches as well as the connection with the traditional scientific disciplines is needed for augmented requirement for fibrous supports (yarns) possessing multifunctionality [196,

151, 46]. Among the first commercial applications in textiles domain, finishing processes using nanoparticles recorded outstanding results. Neverthless, due to the poor fixing of these nanoparticles on the textile surface, these finishes were not resistant to washing. Durable nanofinishings with improved properties in fibrous matrices that are attributed to strong structural bounds between organic macromolecules and inorganic crystals have been reported when functional polymer matrices as host molecues for nanoparticles were used [25, 162]. A simple explanation is that the major part of organic materials is intercrystalline, and in a minor way intracrystalline, meaning improved bonding imparting desired wettability with different functional properties like self-cleaning, photocatalytic, UV resistance, and flame retardancy which are unique characteristics provided by different nanoparticles. For these type of applications, in order the nanoparticles to be dispersed homogeneously in the different matrices, prevent particles agglomeration, and increase the stability of ZnO nanoparticles dispersions, a number of new synthetic strategies have been developed [54, 88, 194, 95, 170, 171]. β-CD contains a toroidal hydrophobic cavity, which is capable of including a variety of inorganic and organic guest species, being a cyclic oligosaccharide consisting of seven α-1,4 linked D-glucopyranose units, [11,

167]. The assistance of *β-cyclodextrin* (*i.e.*MCT- MonoChloroTriazinyl-β-cyclodextrin) in the ZnO synthesis can ensure formation of different morphologies (micro-pellets, rods, plates, boxes, wires, tripods, tetrapods, nodular, irregular) of ZnO primary nanoparticles possessing different levels of zinc/oxygen concentrations and also different defect species dominancy [155]. Presently, CDs are used as nanoreactors, nanoprotectors, nanocapsules, drug releasing agents at a nanometric level [50, 127, 99].

The treatment of fibrous polymeric matrices with CDs has been scarcely reported in literature. Among them, MCT is a biocompatible polymer being known to form a wide range of inclusion complexes with several molecules [12, 40].

Nanoparticles are known to have high surface energy, which is making them extremely reactive, and most systems undergo aggregation without protection of their surfaces. To eliminate or minimize generated waste and implement sustainable processes, green chemistry and chemical processes have been recognized the best paths for the preparation of nanoparticles [136]. As protecting agents of nanoparticles, much effort is now being focused on polysaccharides. For silver nanoparticle growth, soluble starch has been selected as the stabilizing agent and a-D-glucose as the reducing agent in aqueous solution of $AgNO_3$. In order to maintain noble metal (platinum, palladium and silver) nanoparticles in colloid suspension, arabinogalactan has been used as a novel protecting agent [110]. Synthesized platinum, palladium and silver nanoparticles with narrow size distribution have been achieved by using porous cellulose fibers as the stabilizer [58]. The carbonization of cellulose is catalyzed by Pt nanoparticles and mesoporous amorphous carbon is fabricated in high yields. By self-supporting macroporous sponges of silver, gold and copper oxide, as well as composites of silver/copper oxide or silver/titania can be routinely prepared by heating metal–salt-containing pastes of dextran, chosen as a soft template, carbon-based functional composites with metal nanoparticles are obtained [59, 192]. Nanoparticles of metal oxide and sulfides have been prepared employing polysaccharides as stabilizer.

It should be mentioned that degradation of polymeric materials is caused by exposure to various factors such as heat, UV light, irradiation ozone, mechanical stress and microbes. Oxygen, humidity and strain are the promoters of this kind of degradation the consequences being flaws as brittleness, cracking, and fading [30].

Zeta Potential measurement could be performed in order to evaluate the treatment's quality, in terms of stability, [96]. The main role in stabilizing the colloidal solutions is played by the surface electric charges of suspended particles, which ensure the electrostatic rejection and prevents clashes and particle aggregation [90]. Both the degree of rejection-namely the relative stability of the colloidal solutions and the speed at which the particles will travel towards the respective electrode, meaning “electrophoretic mobility” in the electric field will be determined by the size of the charge [23]. Electrokinetic potential – ξ is the potential which appears at the relative movement of the phases in contact [55].

The general values, placed between the limits of (+ 30mV) and (- 30mV), differentiate the stable system from the unstable one. Variation intervals of Zeta Potential and the stability generated by them are:

± 0 [mV] ± 5 [mV] – rapid coagulation;
± 10 [mV] ±30 [mV] – incipient instability;
± 30 [mV] ± 40[mV] – moderate stability;
± 40 [mV] ± 60 [mV] – good stability;
± below 60 [mV] – excellent stability.

Information concerning: the composition of the electrolytic solution, the physical-chemical structure of the solid surface, interactions between the components of the solid surface and the liquid, can be offered by the value and sign of the Zeta Potential [213, 214].

In: Novel Approach of Added-Value Zinc Oxide ... ISBN: 978-1-63483-432-2
Editors: N. Vrinceanu and P. Postolache © 2015 Nova Science Publishers, Inc.

Chapter 1

EXPERIMENTAL PROCEDURES

Narcisa Vrinceanu[1,*], PhD,
Dănuţ Dumitru Dumitraşcu[1], PhD,
Eveline Popovici[2], PhD,
Mirela Petruta Suchea[3], PhD, Gianina Broasca[4], PhD,
Nicoleta Dumitrascu[2], PhD, Gabriela Borcia[2], PhD,
Claudia Mihaela Hristodor[2], PhD,
Diana Ecaterina Tanasa[2], PhD,
Alexandra Nistor (Bărgan)[5], PhD,
Diana Coman[1], PhD, Mihai Ciocoiu[4], PhD,
Daniela Farima[4], PhD, and Cristina Rimbu[6], PhD

[1]"Lucian Blaga" University of Sibiu, Romania
"Al.I.Cuza"University of Iasi, Romania
[2]"Al.I.Cuza"University of Iasi, Romania
[3]"Al. I. Cuza" University of Iasi,
Center of Materials Technology and Laser, School of Applied Technology, Electrical Engineering Department, Technological Educational Institute of Crete, Heraklion, Greece
[4]"Gheorghe Asachi" Technical University, Iasi, Romania
[5]"Petru Poni" Institute of Macromolecular Chemistry, Iasi, Romania
[6]University of Agricultural Sciences and Veterinary Medicine of Iasi, Romania

[*] Corresponding author address. Email: vrinceanu.narcisai@ulbsibiu.ro.

1. Selection of Materials

Using hydrophobic/hydrophilic functional polymer fibrous matrices as dispersion medium for nanoparticles will result in nanofinishings with improved bonding properties in fibrous matrices and will impart desired wettability.

The objectives of the research were to evaluate the direct involvement in the relationship of the morphological, structural and chemical attributes, water vapor sorption-desorption and thermal behavior. Consequently the criteria of the matrix materials choice in the synthesis of the ZnO architectures were dictated by the assistance of two surfactants CTAB (Cetyl Trimethy-lammonium Bromide) and Pluronic P123 and the synergic effect of another two main components, montmorillonite (MMT) and Mono-Chloro-Triazinyl-β-cyclodextrin (MCT).

The selection of matrix materials was made in order to perform the best adhesion between the ZnO nanoparticles and polymer. Consequently, the matrix materials were linen fibrous substrates, in other words – cellulosic polymeric matrices - previously grafted with MCT-β-CD (Mono-Chloro-Triazinyl–β-Cyclodextrin). MCT-β-CD is a commercially available β-cyclodextrin with a reactive monochlorotriazinyl group, used as stabilizer [182, 100]. The so-called *anchor group* which reacts with cellulose hydroxyl radicals and cyclodextrin molecule is covalently bonded to the polymer's surface. The stable bound of cyclodextrin onto the polymeric matrix allows its properties to become intrinsic to the modified supports, thus a new generation of *intelligent textiles* possessing enhanced sorption abilities/capacities, as well as active molecules release was born. Moreover, as polysaccharide, MCT-β-CD shows interesting dynamic supramolecular associations facilitated both by inter- and intra-molecular hydrogen bonding, and by polar groups. When a material is exposed to environmental water vapors, the water molecules firstly react with surface polar groups, forming a molecular monolayer.

MCT-β-CD was also chosen in order to enhance the thermal stability and degradation mechanism of ZnO nanocoated linen fibrous samples. Its effect was catalyzed by montmorillonite (MMT) emulsion was applied on the supports, prior to the ZnO coating.

Montmorillonite (MMT) is a type of inorganic natural clay, which has silicate (SiO_4) tetrahedral sheets arranged into a two-dimensional network structure [37]. MMT can provide thermal resistance, wrinkle resistance and antibacterial properties on textiles [26, 123, 163, 101, and 198].

The effect of the same polymer matrix – meaning the cyclic oligosaccharide, β-cyclodextrin (*i.e.* MCT- *MonoChloroTriazinyl-β-cyclodextrin*), was relevant in the architectural structure and stability/stabilization of strong textures consisting of ZnO nanoparticles dispersions. The assistance of MCT at the ZnO nucleation ensured formation of micro-pellets with nanometric thickness. MCT may also have influence in the shape of the crystal during its growth, the texture being in accordance with the morphology of the single crystal elements. In other words, the fundamental role played by polymer involving in the synthesis in the patterning/architecture/designing of ZnO composites has been stressed. Consequently, by a proper selection of the matrix, the lamellar structure/morphology of obtained architectures, similar to biogenic calcites in the presence of echinoderm intracrystalline proteins, proved the biomimetic, biogenic, and naturally assembled aspect.

Now, regarding the choice of polymeric matrices in our work, it should be stated that, due to their good physical and chemical properties, synthetic fibers are widely used in apparel and home furnishings. Due to its attributes, such as reduced tendency to shrinkage, strength and contractibility, polyester (PES) matrix is one of the most used synthetic fibers. The demand of PES such as (polyethylene terephthalate) in textile market for high performance applications in smart textiles, technical textiles, operation clothing etc. and more recently, for their potential applications as electronic textiles is connected with the mechanical properties, relative high melting point and glass transition temperature (Tg), insensitivity to common solvents and moisture, chemical inertness. Known as a thermoplastic resin obtained by esterification of benzene, 1,4-dicarboxylic acid with 1,2-ethandiol, the ester groups are mainly involved in intermolecular interaction being responsible for the crystallinity, thus PES having a rather hydrophobic responsivity. In addition, ZnO modification of PES matrix, adds value by the benefits of features ranging from detoxification, disinfection, heat and UV protection to biocompatibility and bioactivity. Due to their adequate physical-mechanical characteristics and high endurance to light, woven materials made of polyester fibers were selected as matrix materials.

We also select an inorganic–organic matrix material, based on ZnO microparticles impregnation of polyester fibrous polymeric matrix, to perform combined UV-protection and antimicrobial properties, as well as high hydrophobicity, the choice being based on the real fact that the textile industry demanded development of multifunctional polymeric matrices adapted in particular to UV radiation.

2. NANOMATERIALS

The MCT-β-CD (monochlorotriazinyl–βeta-cyclodextrin) under the trade name CAVATEX or CAVASOL® W7 MCT (CAVATEX), does not irritate nor does initiate sensitization. It is an effective tool to modify textile surfaces, purchased by the company Wacker Chemie AG [137].

$Zn(OAc)_2$, with an assay of 97%, was obtained from the Romanian Chemical Company - CHIMOPAR. CTAB was provided by Merck Company, with an assay of 97%. Urea was supplied by CHIMOPAR, Bucharest.

Acetic acid (assay 99%) was purchased from Chemical Company Society, Romania.

Bentonite montmorillonitic-clay (MMT), with a remarkable compositional complexity, was provided by Riedel-de Haen Chemicals Company. As a first step, in order to reach the transition cations sodium form, it was necessary to perform an ion exchange process, for their cleansing. A 1M NaCl solution with a ratio solid/liquid 1:10 was employed, for the clay exchange in Na^+ form.

From CLARIANT two agents were purchased, meaning: a nonionic dispersing agent, SETAVIN and a binder agent, Appretan TT.FR liq from CLARIANT 10% wt.

A binder resin was obtained from ARALDITE Company.

The linen fabric, two 100% twill linen supports, each of size 3 cm × 3 cm, was kindly provided by Iasi Tex Company, Romania. The supports had been resized, scoured and bleached.

Woven 100% polyester fibrous matrices (97.5 g/m2) were donated by SC Condor SA, Romania, selected and prepared for impregnation in the form of 5 cm ×10 cm strips. The woven (polyethylene terephthalate) complex fabric has also the following characteristics, presented in Table 1: (2Weft+1Warp); F:Diagonal 2/1 Z:R; canvas, thickness 0.23 mm. Table 1 displayed a more detailed description of the characteristics belonging to PES polymeric matrix.

Table 1. The main characteristics of the studied textiles

Composition of the woven fabric	Structure of the woven fabric	Density of Tex length		Rupture strength [N]		Relative elongation [%]		Absolute elongation [mm]		Thickness [mm]
		Warp	Weft	Warp	Weft	Warp	Weft	Warp	Weft	
100% PES	Diagonal 2/1	11	18	671.6	717.4	27.494	30.44	33.07	33.56	0.23

ZnO powder which has micrometer-size particles, an assay of 99,5% and a molecular mass of 81.38 g, is supplied from Lach-ner. The particle-size distribution, determined by laser diffraction, shows that the composition of the powder consists of 1.7 μm size particles (75% proportion). As active agent, zinc oxide (ZnO) was dispersed in an aqueous solution of commercially available acrylic copolymer Apretan from Clariant Company.

3. Nanofabrication Fundamental Techniques for Synthesizing

This part consists of two subdivisions. The first subdivision is dedicated to synthesis of nano-scaled ZnO and the polymeric matrix (cellulose) it has been encaved in.

3.1. Synthesis of ZnO Nanoparticles and Their Application on the Same Type of Polymeric Matrices, Meaning Cellulosic Fibrous Matrices (Linen Supports)

3.1.1. In-Situ Synthesis of Nano-ZnO Particles on Cellulosic Fibrous Matrix (Linen Fibrous Support), with the Assistance of Two Surfactants

For the investigation of humidity adsorptive responsivity of the resulted composites, the procedure was as follows. The research used the fibrous polymeric matrices (yarns) previously grafted/functionalized with MCT-β-CD. The grafting process of the textile fabric was performed following two other processes: the exhaustion and squeezing treatment and the heat treatment at 160°C [48].

ZnO nanoparticles were synthesized *in-situ* by using the hydrothermal method on fibrous polymeric matrices (yarns) having a certain concentration of MCT-β-CD. The samples with sizes of 30 x 30 cm^2 were immersed in the solution prepared as follows: to form a uniform solution, dissolve zinc acetate $Zn(CH_3COO)_2$. $2H_2O$, purity – 99%) (0.005 mol/1000mL) as precursor, in deionized water by stirring. Add dropwisely 0,1 mol of urea solution with constant stirring. Add dropwisely acetic acid to adjust to 5 the pH value of the mixed solution and vigorously stir the final reaction mixture for two hours at room temperature. Pour the final mixture into 100 mL stainless-steel autoclaves made of Teflon (polytetrafluoroethylene). Immerse the fibrous polymeric matrices (yarns) and place the autoclaves in the oven for the hydrothermal treatment at 90℃ overnight. Cool the autoclaves down to room temperature, remove the matrices from the autoclaves, wash of samples several times with distilled water and dry at 60°C overnight for complete conversion of the remaining zinc hydroxide to zinc oxide (Figure 1).

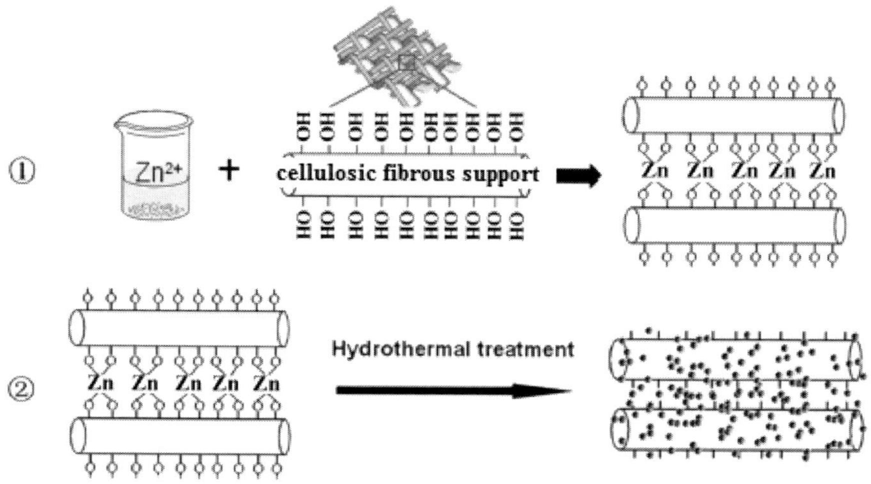

Figure 1. Flow chart for the preparation methodology of ZnO nanoparticles coated polymeric (cellulosic) fibrous matrix.

3.1.2. Deposition of ZnO nanoparticles on cellulosic fibrous matrices with the assistance of montmorilonitic clay and Mono-chloro-triazinyl-B – Cyclodextrin

To make composite samples, the particle sizes of the montmorillonite (MMT) were reduced by an ultrasonic crashing machine. In order to prepare a good MMT emulsion, so that that the tiny particles of MMT should be well dispersed and will not be aggregated again to form large molecules, a nonionic dispersing agent was used. The emulsion was prepared as follows:

- Add 1g of MMT into 42 mL of dispersing agent;
- Stir magnetically stirring the emulsion for 2 hours;
- Crash ultrasonically the emulsion for 30 minutes, to reduce the particle sizes;
- Monitor the behavior of the emulsion at different time intervals, i.e. 5, 10, 15, 30 minutes;
- Prepare the binder resin solution;
- Dropwisely pour the binder resin solution;
- Stir the mixture until homogeneousness;

- Add ZnO nanoparticles emulsion with concentration of 0.5% into the mixture;
- The resulting composite was allowed to stay overnight to remove air bubbles, before use.

Table 2. Synthesis conditions for each of the sample

Sample	**Specifications**
Reference 1	Cellulosic fibrous matrix (linen fibrous support)
Reference 2	ZnO powder hydrothermally synthesized, non-calcinated
Sample 1	Functionalization of cellulosic fibrous matrix (linen support) with MCT- β –CD (Monochlorotriazinyl–β -Cyclodextrin) by exhaustion and thermal treatment
Sample 3	ZnO powder hydrothermally synthesized on cellulosic fibrous matrix (linen fibrous support)
Sample 4	ZnO powder hydrothermally synthesized on functionalized cellulosic fibrous matrix (linen fibrous support)
Sample 5	ZnO powder hydrothermally synthesized on functionalized cellulosic fibrous matrix (linen fibrous support) with the assistance of CTAB (Cetyl Trimethylammonium Bromide)
Sample 6	ZnO powder hydrothermally synthesized on functionalized cellulosic fibrous matrix (linen fibrous support) with the assistance of P123 (**M**ono**c**hloro**t**riazinyl-β – **C**yclo**d**extrin)
Sample 7	MCT- β –CD (**M**ono**C**hloro**T**riazinyl–**β** -**C**yclo**D**extrin)

The synthesis of this oxide by hydrothermal method has been described elsewhere [169].

After preparing the MMT-ZnO emulsion, it was padded onto the four different cellulosic fibrous matrices (linen fibrous supports):

- Reference linen support (without MCT) and ZnO;
- Cellulosic fibrous matrices (linen support) with MCT and without ZnO;
- Non-functionalized cellulosic fibrous matrices (linen support) without MCT and with ZnO;
- Functionalized cellulosic fibrous matrices (linen support) with MCT and ZnO, using a padding machine.

In order to prevent the reduced particles from aggregating again to form larger molecules because of the poor dispersing properties of the MMT particles, the padding process should be carried out as soon as possible, i.e. no longer than 30 minutes after finishing the crashing process. The emulsion was padded onto the cellulosic fibrous matrices (linen support) by the padder with the pressure of 3 kg/cm^2 and the speed of 5 rpm. Two padding times were investigated including three padding times and six padding times. Subsequently the wet pick-up was 84.21% and 88% respectively. Finally, the treated fibrous matrices were dried at 140°C, for 3 minutes.

Washing was carried out to remove the byproducts. The treated supports were immersed for 5 min in 0.2 g/L sodium lauryl sulfate dodecyl, to remove the unbound nanoparticles. Then the cellulosic fibrous matrices (linen support) were rinsed at least 10 times to completely take out all the soap solution. The samples thus washed were air-dried. Simultaneously, a reference emulsion made by MMT without ZnO was applied for comparison.

The thermal treatment consisted of two main stages carried out into the calcination oven. Firstly, the samples were subjected to an increasing of temperature up to 150°C; secondly the probes were heated up to 350°, 450° respectively.

Table 3. Specifications and descriptions of the samples

Samples name	Description
MCT	monochlorotriazinyl–βeta-cyclodextrin
MMT	Bentonite montmorillonitic-clay
ZnO	ZnO powder hydrothermally synthesized
L1	Non-functionalized cellulosic fibrous matrices (linen support) without MCT and with ZnO, with the assistance of MMT, thermally non-treated
L2	Non-functionalized cellulosic fibrous matrices (linen support) without MCT and with ZnO, with the assistance of MMT, treated at 150°C
L3	Non-functionalized cellulosic fibrous matrices (linen support) without MCT and with ZnO, with the assistance of MMT, treated at 350°C
L/MCT-1	Functionalized cellulosic fibrous matrices (linen support) with MCT and ZnO, with the assistance of MMT, thermally non-treated
L/MCT-2	Functionalized cellulosic fibrous matrices (linen support) with MCT and ZnO, with the assistance of MMT, treated at 150°C
L/MCT-3	Functionalized cellulosic fibrous matrices (linen support) with MCT and ZnO, with the assistance of MMT, treated at 450°C

We proceeded to consider for investigation only the last two samples:

- Non-functionalized cellulosic fibrous matrices (linen support) without MCT and with ZnO, with the assistance of MMT;
- Functionalized cellulosic fibrous matrices (linen support) with MCT and ZnO, with the assistance of MMT (Table 2)

The specifications and characteristics of the samples are showed in the above:

3.1.3. Novel nanofabrication of nano-ZnO by pelletization

This part is assigned to the synthesis of zinc oxide (ZnO) textures, without their subsequent deposition onto a substrate previously prepared. A wet chemical method (sol-gel method) [132] was used involving zinc nitrate and sodium hydroxide as precursors and soluble MCT as stabilizing agent.

The methodology is as follows:

- Dissolve various amounts of soluble (MCT) Monochlorotriazinyl-β-cyclodextrin (MCT_1 = 0.5%, MCT_2 = 1%, MCT_3 = and 1.5%, MCT_4 = 2%) in 500 mL of distilled water;
- Add zinc nitrate in the above solutions;
- Magnetic constant stir of the solutions to completely dissolve the zinc nitrate;
- Add of 0.2 mole of sodium hydroxide solution under constant stirring, drop by drop touching the walls of the vessel;
- Allow the reaction to proceed for two hours after complete addition of sodium hydroxide;
- Allow the solution to settle for overnight;
- Carefully discard the supernatant solution;
- Centrifuge the remaining solution at 10000 rot/min for 10 min and discard the supernatant ;
- Wash with distilled water of the solution three times, to remove the byproducts and the excessive MCT that bound with the nanoparticles;
- Dry at 80°C overnight; during drying, a complete conversion of $Zn(OH)_2$ into ZnO takes place.

The second subdivision is tributary to the application of controllable *bulk ZnO architectures onto polyester (PES) fibrous matrices.*

3.2. Preparation of Hybrid Organic Inorganic Composites Polymeric Composites

3.2.1. Preparation of Hybrid Organic Inorganic Composites Polymeric Composites Made of PES Fibrous Matrices Covered/Finished with Bulk/Micro Scaled/Powdered ZnO As Candidates to Antimicrobial Character

The experimental conditions corresponding to the preparation of samples are presented in Table 4.

Table 4. Experimental research versions

Research routes	Composition of the antimicrobial solution
V0	Polyester fibrous matrices+Apretan (80g/l)
V1	Polyester fibrous matrices +Apretan(80g/l) + 1% ZnO
V2	Polyester fibrous matrices +Apretan(80g/l) + 3% ZnO
V3	Polyester fibrous matrices + Apretan(80g/l) + 5% ZnO
V4	Polyester fibrous matrices + Apretan(80g/l) + 7% ZnO

The treatment of polyester fibrous matrices (woven fabrics) with ZnO microparticles dispersed in two different media: water and methanol was performed based on the procedure described in reference [20], as follows:

- Prepare the active (antimicrobial, UV protective, hydrophobic) solutions according to Table 3. Prepare the solutions with ZnO powder dispersed in methanol (99.8%), at 1%, 3%, 5% and 7% concentration. Stir the solutions for 15 min, adding 4 drops of Vitexol (BASF) to avoid foaming and 80 g/l Apretan (Clariant) to ensure bonding of ZnO to fibers.
- Wash, rinse and air dry of fibrous matrices, before coating;
- Impregnate, in laboratory conditions, of polyester fibrous matrices (woven fabrics) with the active solution.
- Pad under laboratory conditions, using Wermer Mathis AG equipment (by pollution, followed by wet-picking, pressing the textile between hydroextraction cylinders, for removal of a part of liquid/paste from the textile surface). Weigh samples before and after impregnation.
- Dry for 3 minutes at 110 °C, under laboratory conditions using a Vetter machine.
- Stabilize via thermal induced cross-linking by curing for 3 minutes at 150 °C, under laboratory conditions using a Vetter machine.

3.2.2. Preparation of Hybrid Organic Inorganic Polymeric Composites Made of PES Fibrous Matrices Covered/Finished with Bulk/Micro Scaled/Powdered ZnO As Candidates to Anti-UV and High Hydrophobic Character

Dispersions with appropriate viscosity of ZnO powder (3%, 5%, 7% wt.) and Appretan (10% wt.) in water and methanol respectively have been prepared. Appretan is a hard, coarse vinyl acetate dispersion used to handle modification of apparel fabrics and knitwear, bonding of nonwoven and wadding, and coating of technical textiles. It has a nonionic character. Binder properties are dependent upon the glass transition temperature (Tg) of the polymer. Due to its low Tg, vinyl acetate is not that stiff, and it has the advantage of low cost.

The deposition of the microstructural coatings, such as the dispersions onto the polymeric (PES) fibrous matrix was made using a special kit – micrometer double blade wet film applicator EQ-Se-KTQ-150D. The thickness of the coating was controlled by the width of the two blades. A width of 0.05 mm was used. Micrometer head accuracy is 10 microns.

The samples were dried and heated at 120°C for 6 minutes.

The following samples were prepared:

Table 5. Specifications of prepared samples

Modification 1	Modification 2	Modification 3
0-PES	0-PES	0-PES
1_ polymeric (PES) fibrous matrix + acrylic binder	6_ polymeric (PES) fibrous matrix + 1% ZnO + methanol	10_ polymeric (PES) fibrous matrix + 1% ZnO + methanol/48 hours
2_ polymeric (PES) fibrous matrix + 1% ZnO	7_ polymeric (PES) fibrous matrix +3%ZnO+methanol	11_ polymeric (PES) fibrous matrix + 3%ZnO + methanol/48 hours
3_ polymeric (PES) fibrous matrix + 3%ZnO	8_ polymeric (PES) fibrous matrix + 5%ZnO + methanol	12_ polymeric (PES) fibrous matrix + 5%ZnO + methanol/48 hours
4_ polymeric (PES) fibrous matrix + 5%ZnO	9_PES+ 7%ZnO + methanol	13_PES + 7%ZnO + methanol/48 hours
5_ polymeric (PES) fibrous matrix + 7%ZnO		

- 100% polymeric (PES) fibrous matrix modified with 10% Appretan solution (for reference);
- 100% polymeric (PES) fibrous matrix modified with dispersion prepared as described above (ZnO concentrations 1%, 3%, 5% and 7%) in water;
- 100% polymeric (PES) fibrous matrix modified with dispersion prepared as described above (ZnO concentrations 1%, 3%, 5% and 7%) in methanol;
- 100% polymeric (PES) fibrous matrix modified with dispersion prepared as described above (ZnO concentrations 1%, 3%, 5% and 7%) in methanol, aged for 48 hours.

Description and denomination of each prepared sample (samples specifications) are presented in Table 5.

4. Instrumentation for Characterization of Composites

The selected methods allow for analysis of the surface and volume properties of the impregnated materials, all related to the responsivity of the fabric to environmental factors, as UV radiation, water and water vapors, under controlled laboratory conditions.

The instrumental methods were conducted to measure the particle sizes of the ZnO powder in order to characterize the surface morphology and chemical composition of the treated supports.

Scanning Electron Microscopy (SEM) - And Energy Dispersive X-Ray Spectroscopy (EDX)

The scanning Electron Microscope (SEM) images was carried out with a Quanta 200 3-D Dual Beam type microscope, from FEI Holland, coupled with an EDS analysis system manufactured by EDAX - AMETEK Holland equipped with a SDD type detector (silicon drift detector) with a magnification of 300X, 1200X, respectively. Taking into account the sample type, the analyses were performed, using Low Vacuum working mode, allowing the probes testing in their initial state, without a previous metallization (as in High Vacuum working type).

Both for the acquisition of secondary electrons images (SE – secondary electrons) and EDS type elemental chemical analyses, LFD (Large Field Detector) type detector was used, running at a pressure of 60 Pa, and a voltage.

Atomic Force Microscopy (AFM)

Atomic force microscopy was performed on the composite ZnO-PES fibers using a system Digital Instruments NanoScope IIIa SPM from Veeco, in Contact Mode. Local roughness of ZnO onto the fiber surface was estimated using The NanoScope real-time control software allowing scan parameters to be changed without stopping and restarting scans. Cross-section analysis was performed using off-line Surface Profile Analysis mode NanoScope software.

X-Ray Diffractometry

The samples were tightly packed into the sample holder. X-ray Diffraction (XRD) data for structural characterization of the various prepared samples were collected on an X-ray diffractometer (PW1710) using Cu-Kα radiation (k = 1.54 Å) source (applied voltage 40 kV, current 40 mA). About 0.5 g of the dried particles were deposited as a randomly oriented powder onto a Plexiglass sample container, and the XRD patterns were recorded at 2θ angles between 20 ° and 80°, with a scan rate of 1.5°/min. Radiation was detected with a proportional detector.

Evaluation of Crystallinity

The extent of crystallinity (*Ic*) was estimated by means of Eq. (1), I_{020} is the intensity of the 020 diffraction peak at 2θ angle close to 22.6°, representing the crystalline region of the material, and I_{am} is the minimum between 200 and 110 peaks at 2θ angle close to 18°, representing the amorphous region of the material in cellulose fibers [137, 117, 124]. I_{020} represents both crystalline and amorphous materials while I_{am} represents the amorphous material.

$$I_C = \frac{I_{020} - I_{am}}{I_{020}} x100(\%) \tag{1}$$

A *shape factor* is used in x-ray diffraction to correlate the size of sub-micrometer particles, or crystallites, in a solid to the broadening of a peak in a diffraction pattern. In the Scherrer equation,

$$\tau = \frac{K \bullet \lambda}{\beta \cos\theta}$$

where K is the shape factor, λ is the x-ray wavelength, β is the line broadening at half the maximum intensity (FWHM) in radians, and θ is the Bragg angle [137, 117, 124]. τ is the mean size of the ordered (crystalline) domains, which may be smaller or equal to the grain size. The dimensionless shape factor has a typical value of about 0.9, but varies with the actual shape of the crystallite. The Scherrer equation is limited to nano-scale particles.

FT-IR Spectroscopy

FT-IR was used to examine changes in the molecular structures of the samples. Analysis has been recorded on a FT-IR JASCO 660+ spectrometer Fourier Transform Infrared Spectrometer operating in the 4000-450 cm^{-1} range. The analysis of studied samples was performed at 2 cm^{-1} resolution in transmission mode. Typically, 64 scans were signal averaged to reduce spectral noise. The KBr pellet technique was used to prepare the powder samples for IR studies. Film samples were powdered, mixed with KBr, and then pressed into pellets before recording the spectra. The average of three scans for each sample was taken for the peak identification.

Study of Humidity (Water Vapors) Sorption/Specific Surface Area

Dynamic vapors sorption (DVS) capacity for the studied samples, at 25°C averaging in the domain of relative humidity (RH) 0-90% has been investigated by using an IGAsorp apparatus, a fully automated gravimetric analyzer, supplied by Hiden Analytical, Warrington - UK). It is a standard sorption equipment, which has a sensitive microbalance (resolution 1µg and capacity 200 mg), which continuously registers the weight of the sample in terms of relative humidity change, at a temperature kept constant by means of a thermostatically controlled water bath. The measuring system is controlled by appropriate software.

This apparatus is used to study water sorption at atmospheric pressure by passing a humidified stream of gas over the sample.

Thermal Analysis

The differential scanning calorimetry analysis (DSC) of fibrous supports-ZnO composites was carried out using a NETZSCH DSC 200 F3 MAIA instrument under nitrogen. Initial sample weight was set as 30-50 mg for each operation. The specimen was heated from room temperature to 350°C at a heating rate of 10°C/min.

Thermogravimetric Measurements

Thermogravimetric analysis (TGA) was performed using a MOM Q-1500D System from Paulik-Erdey Company (Hungary) in nitrogen environment, in the temperature range 25–750 °C, using a heating rate of 10° C/min.

Air Permeability and Porosity Measurements

Air permeability is related to the rate of airflow passing perpendicularly through a known area under a set air pressure between the two surfaces of a material.

The rate of airflow passing perpendicularly through a known area of fabric was adjusted to obtain a set differential air pressure between the two fabric surfaces. From this rate of airflow, the air permeability of the fabric was determined. Air permeability was measured using a TEXTEST Fx 3300 permeometer, using a pressure differential of 12.7 mm of water as specified in ASTM D737. The resistance to vapor transfer is evaluated according to ASTM D1518, by guarded hot plate method.

The formula used for air permeability estimation was:

$$P_A = \frac{V}{t \times A}(l / m^2 \times s)$$

where:

V – air volume passing a textile interface, a pressure difference is created in between;

t – time of air passing [s];

A – the surface of sample subjected to analysis [m^2].

Pore size was measured as specified in ASTM E1294-89 using a porometer. ASTM E1294-89 is a liquid extrusion method of analysis for measuring the porosity of textile materials.

For porosity determination, tests were carried out using dry and wet samples. The specimen was wetted first with Silwick (surface tension 19.8 dynes/cm) a liquid of low surface tension, and then extruded under increased air pressure. The maximum pore size was determined in correlation with the first airflow, which was considered as the bubble point. As smaller pores were emptied, the airflow through the sample increases, and it was recorded as a function of air permeability. All these results were estimated in comparison with that of the dry sample.

Sample performance in water vapors passing resistance, was measured according to ISO 811: a specimen was subjected to a steadily increasing pressure of water on one face, under standard conditions, until penetration occurs in three places. The water pressure may be applied from below or from above the test specimen. The hydrostatic head supported by a fabric is a measure of the resistance to the passage of water through the fabric.

The Determination of Zeta Potential

The distribution of ions in the surrounding interfacial region is affected by the development of a net charge at the particle surface, resulting in an increased concentration of counter ions (ions of opposite charge to that of the particle) close to the surface. Thus around each particle, an electrical double layer exists. The liquid layer surrounding the particle exists as two parts; an inner region, called the Stern layer, where the ions are strongly bound and an outer, diffuse, region where they are less firmly attached. Within the diffuse layer, the ions and particles form a stable entity inside a notional boundary.

When a particle moves (e.g. due to gravity), ions within the boundary move with it, but any ions beyond the boundary do not travel with the particle. This boundary is called the surface of hydrodynamic shear or slipping plane. The potential that exists at this boundary is known as the *Zeta potential.*

The potential stability of the colloidal system is indicated by the magnitude of the zeta potential. A colloidal system is when one of the three states of matter: gas, liquid and solid, are finely dispersed in one of the others. If all the particles in suspension have a large negative or positive zeta potential then they will tend to repel each other and there is no tendency to flocculate. However, if the particles have low zeta potential values then there is no force to prevent the particles coming together and flocculating. The general dividing line between stable and unstable suspensions is generally taken at either + 30mV or -30mV. Particles with zeta potentials more positive than +30mV or more negative than -30mV are normally considered stable. The most important factor that affects zeta potential is pH. A zeta potential value on its own without a quoted pH is a virtually meaningless number.

A consequence of the existence of electrical charges on the surface of particles is that they will exhibit certain effects under the influence of an applied electric field. These effects are collectively defined as electrokinetic effects. Depending on the way in which the motion is induced, there are four distinct effects. These are:

- Electrophoresis: the movement of a charged particle relative to the liquid it is suspended under the influence of an applied electric field.
- Electro-osmosis: the movement of a liquid relative to a stationary charged surface under the influence of an electric field.
- Streaming potential: the electric field generated when a liquid is forced to flow past a stationary charged surface.
- Sedimentation potential: the electric field generated when charged particles move relative to a stationary liquid.

The employing of streaming potential has already been reported by various studies, in order to characterize the electrokinetic potential of the interior pore surfaces in microporous membranes, [174, 83, 65, 116].

The principles of streaming potential experimental evaluation consist in the following: an electrolyte solution is forced, through hydraulic pressure, to flow through a porous plug of material, across a channel formed by two plates, or down a capillary; thus a streaming potential is generated. The liquid in the channel carries a net charge. Due to hydraulic pressure, its flow gives rise to a streaming current, thereby generating a potential difference. The result is a mechanical transfer of charge opposed by this potential, causing back conduction by ion diffusion and electro-osmotic flow (due to the potential difference). Once equilibrium condition is attained, the leak current is cancelled by the streaming current, and measured potential difference is the streaming potential [159, 64].

The relationship between streaming potential and zeta potential is given by Helmholtz-Smoluchowski equation [67, 68]:

$$\xi \equiv \frac{E_S}{\Delta P} \times \frac{\eta}{\varepsilon \varepsilon_0} \times \frac{L}{A} \times \frac{1}{R}$$

Zeta potential evaluation and streaming potential measurements for the samples included in this thesis were carried out using a streaming potential analyzer (ZetaCad from CAD Instrumentation Company). The main parts of the instrument are the analyzer, the data control system, and the measuring cells. The analyzer consists of a mechanical drive unit producing and measuring the pressure that drives the electrolyte solution from a reservoir into and through the measuring cell. A pump with a flow capacity of 1.3 L/min achieves the circulation of the solution. The instrument measures the streaming potential, as well as the streaming current, simultaneously.

Two different measuring cells can be used: cylindrical or rectangular. Depending on the nature of the sample under investigation, the choice of which cell to use is made. A cylindrical cell is used to measure the zeta potential of solid bodies such as fibers, granules, and rods. The samples presented in this book were analyzed using the rectangular cell. This cell is a PMMA cell having 125x50 mm dimensions. A channel of known size (75 x 9.5 mm) is created by the use of PTFE spacers. KCl electrodes that measure the induced streaming potential are mounted at each end of the channel. Jacobasch and Schurz gave schematic illustrations of the electrokinetic analyzer and associated measuring cells [68].

Measurements of zeta potential of the antimicrobial treated polymeric (PES) fibrous matrix (woven fabrics) (Table 3), before and after washing, were performed using an electrolyte solution obtained by mixing 1800 mL distilled water with 20 mL KCl 10^{-1}M similar to previously presented reports [166] at different values of pH (4, 7, 9) [19].

After samples immersion into electrolyte solution for two hours, Zeta Potential measurements were performed. In order to quantify the time stability of the treatment with ZnO, specimens were washed at 40 °C, using Washteck laboratory machine, followed by room temperature drying.

Physical-mechanical properties are evaluated for uncoated and coated textiles. *The tensile strength* is calculated according to ISO 2062, using a H5K-T dynamometer equipped with QMAT TEXTILE software.

The UV reflectance determines the photoprotective performance of the material, in relation to its physical–chemical properties. In this respect, the absolute reflectance, in the UV range, is measured using an Avantes optical system, consisting of an AvaSpec-2048-USB2 spectrometer, an AvaLight-DHc Compact deuterium-halogen light source, an UV/VIS reflection probe and a recording system.

Impedance Spectroscopy determines *the dielectric properties* of the textile. The polymeric matrices are dielectric materials, prone to accumulate electrostatic charge, which affects the garment quality and comfort to human contact and conducts to attracting particles in suspension in environmental air. In relation to the dielectric properties, ZnO coating is modifying the charge effect on textiles. Because it takes into account the reactive, capacitive and resistive character of the polymeric matrix, also the phase difference, as response to an applied AC electrical field, the impedance is a more general concept than the resistance.

The evaluation of the dielectric character can be performed using an AC electrical field with frequencies in a wide range of values, where the response of the sample is due to conduction and polarization. The conductive character of the polyester fibrous matrix is expected to be very limited, since common polymers are insulators. Taking into account that the polarization builds up by orientation of the polar groups (permanent dipoles) in the polymeric chains and induced dipoles through charge displacement, the dielectric character relates to the polarization process. In this way, a different dielectric behavior can be identified at low and high frequency, as the polar groups follow the AC field only at low frequency.

The impedance Z: $Z(\omega)=Z' + jZ''$

is plotted in rectangular coordinates, as $Re(Z) \equiv Z' = |Z| x \cos\theta$ and $Im(Z) \equiv Z'' = |Z| x \sin\theta$, where Z' and Z" characterize the resistive and capacitive behavior of the samples, respectively, on the Cole–Cole diagrams.

$$Z'' = \frac{1}{\omega C} \qquad (2)$$

where C is the capacitance, $\omega = 2\pi f$ the ac angular frequency and $\theta = \tan^{-1}(Z''/Z')$ the phase angle.

By using the electrodynamics equation, the AC conductivity for the three polymeric matrices is compared:

$$\varepsilon''(f) = \frac{\sigma(f)}{2\pi f \varepsilon_0} \qquad (3)$$

where ε" is the imaginary part of the electrical permittivity, σ is the conductivity and f is the frequency of the AC electric field. Eq. (3) allows computing the conductivity, as a function of frequency, from the relaxation of ε".

In order to measure the impedance a Solatron 1260A Impedance Spectroscopy system, an electrode kit with two plates with adjustable thickness and an AC electrical field 1 mV in amplitude, 101–106 Hz frequency range are used. All measurements are made at controlled room temperature (22°C) and humidity (60%), keeping the same thickness of samples, following an initial stabilization of working parameters. The samples are 2 cm × 2 cm in size.

The impedance data are determined for the reference material (not impregnated polyester fibrous matrix) and on the impregnated matrices, before and after UV irradiation, taking into account that the UV exposure could shift the charge effect on their surface. Using a 100 W deuterium lamp, UV irradiation is carried out on both sides of the samples, for 4 h exposure each. In order to test the effect of UV and the stability with time of the material, the measurement is performed at various intervals after irradiation.

Wettability means the ability of a liquid to displace a gas from a solid surface [211, 201, 92, 71, 191, 27]. In terms of behavior in the presence of water, the surfaces can be divided into two main groups: hydrophobic and hydrophilic. The wettability of the surface is estimated by contact angle (CA) measurement. A CA of less than 5° means superhydrophilic surface while a CA between 5° and 90° means hydrophilic surface. In case of hydrophobic surface, CA values ranging from 90° to 149° means a so-called hydrophobic surface, and a CA of more than 150° is known as superhydrophobic surface [160, 184, 63, 114, 105,]. To achieve a superhydrophobic surface, two methods are usually used: roughening the hydrophobic surface and coating with low surfacc energy materials [73] various types of fibrous matrices, as these textured materials are not usually compact and flat enough to allow a realistic contact angle to be measured. In addition, they may also absorb the liquid drops.

The contact angles are obtained using:

a. The sessile drop technique, under controlled conditions of room temperature and humidity. An automated system is used to store the drop images, via a Canon A85 camera, with PC-based control, acquisition and data processing. The values of the static contact angle presented are the average of ten measured values obtained on the imaged sessile liquid drop profile, for 1 μl drop size. Bidistilled water is used as test liquid. The adhesion work of water Wa on the surface, which relates to the wettability, is calculated as [66]:

$$\mathrm{Wa} = \gamma_{lv}(1+\cos\theta)$$

where θ is the contact angle and γ_{lv} is the surface tension $\gamma_{lv} = 72.8$ mJ/m^2.

b. Dynamic Contact Angle (DCA) can be defined when the three phase line is in controlled motion. The Dynamic Contact Angles are divided into Advancing and Receding contact angles, meaning contact angles measured when the three-phase line is in controlled movement the liquid wetting the solid or by withdrawing the liquid from a pre-wetted surface, respectively [128]. In the case of chemical heterogenity the surface has domains with different contact angles. For example when wetted with water, hydrophobic domains will pin the motion of the contact line as the liquid advances thus increasing the contact angles. When the water recedes, the hydrophilic domains will hold back the draining motion of the contact line, thus decreasing the contact angle. From this analysis it can concluded that, when tested with water, advancing angles will be sensitive to the hydrophobic domains and receding angles will characterize the hydrophilic domains on the surface. Dynamic contact angles were measured for all the samples prepared as described above. Water was used as measuring liquid. In this experiment, was found out that surface roughness is only one of the main reasons causing the CA values modification, especially for the receding contact angle. In fact, the water molecules should overcome the forces of the chemical bonds when adsorbing or leaving the surface of fibers. Contact angle measurements were performed using a modular high performance computer controlled instrument (KSV Sigma 700). Employing the latest advances in precision microbalance technology was used for the surface properties evaluation both in solution and in solid state. The equipment includes surface tension, critical micelle concentration (CMC), dynamic contact angle (DCA) and powder wettability modules. Dynamic contact angles were measured for all the samples prepared as described elsewhere. Water and ethylene glycol (EG) were used as measure liquids. The DCA runs were performed on rectangular films of about 10 x 0.15 mm.

In: Novel Approach of Added-Value Zinc Oxide ... ISBN: 978-1-63483-432-2
Editors: N. Vrinceanu and P. Postolache © 2015 Nova Science Publishers, Inc.

Chapter 2

RESULTS AND DISCUSSION: ANALYSES AND DISCOVERIES

Narcisa Vrinceanu[1*], PhD, Valentin Petrescu[1], PhD Toderița Nemeș[1], PhD, Eveline Popovici[2], PhD, Mirela Petruta Suchea[2], PhD, Gianina Broasca[3], PhD Nicoleta Dumitrascu[2], PhD, Gabriela Borcia[2], PhD Claudia Mihaela Hristodor[2], PhD, Diana Ecaterina Tanasa[2], PhD, Alexandra Nistor[4], PhD, Diana Coman[1], PhD, Mihai Ciocoiu[5], PhD, Daniela Farima[5], PhD, and Cristina Rimbu[6], PhD

[1]"Lucian Blaga" University of Sibiu, Romania,
"Al.I.Cuza"University of Iasi, Romania
[2]"Al. I. Cuza" University of Iasi
Center of Materials Technology and Laser, School of Applied Technology,
Electrical Engineering Department,
Technological Educational Institute of Crete, Heraklion, Greece
[3]"Gheorghe Asachi" Technical University, Iasi, Romania
[4]Petru Poni" Institute of Macromolecular Chemistry, Iasi, Romania
[5]„Gheorghe Asachi" Technical University, Iasi, Romania
[6]University of Agricultural Sciences and Veterinary Medicine of Iasi, Romania

[*] Corresponding author address. Email: vrinceanu.narcisai@ulbsibiu.ro.

II.1. Morphology Diversity of Polymeric Fibrous Matrices Covered with Bulk/Micro and Nano ZnO

II.1.1. Morphological Profile of Hybrid Organic- Inorganic Materials Made from *In-Situ* Synthesized Nano-ZnO Particles on Cellulosic Fibrous Matrices (Linen Supports), with the Assistance of CTAB and P123

SEM images below represent Reference 1 of initial polymeric fibrous matrix (linen support) (Figure 2a), Reference 2 of ZnO powder (Figure 2b) and Sample 1, under different magnifications.

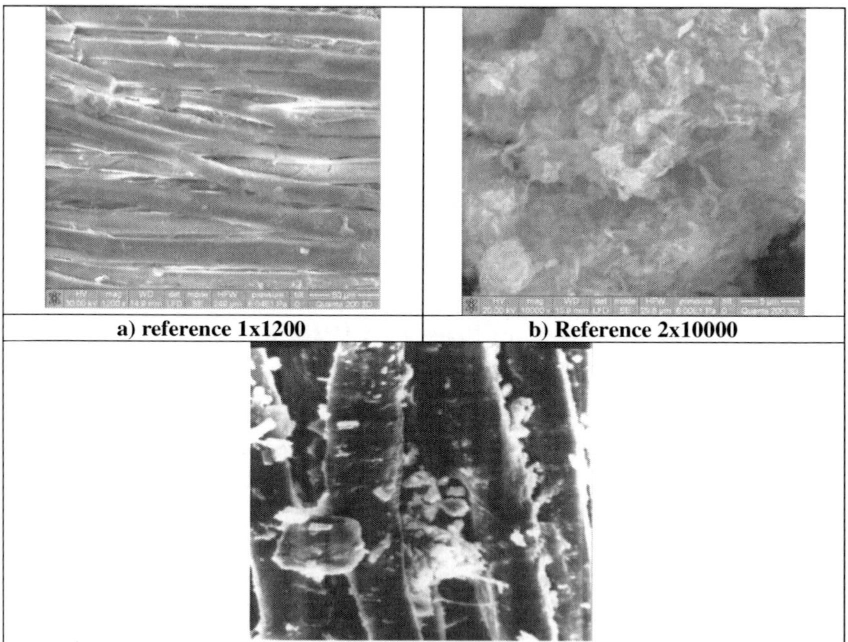

a) reference 1x1200 **b) Reference 2x10000**

c)Sample 1x1200

Figure 2. SEM images of the reference samples and the functionalized polymeric fibrous matrix (linen support) with MCT- β –CD sample.

Images clearly reveal distinguished hexagonal shape of ZnO agglomerations and the morphology of linen fibers with clean behavior (Figure

2c). The SEM images of functionalized polymeric fibrous matrices (linen supports) coated with ZnO (Figure 2a and b) indicate different shapes of deposited ZnO, according to the types of used surfactant. Therefore, ZnO nanoparticles exhibited hexagonal behavior like flowers of ZnO nanocrystals, if the treatment was assisted by P123 surfactant (Figure 4.2a) and lamellar morphology if the treatment was assisted by CTAB (Figure 2b) respectively.

An even cover of particles can be seen on the polymeric fibrous matrix surface and as results, the last one became coarser after the treatment.

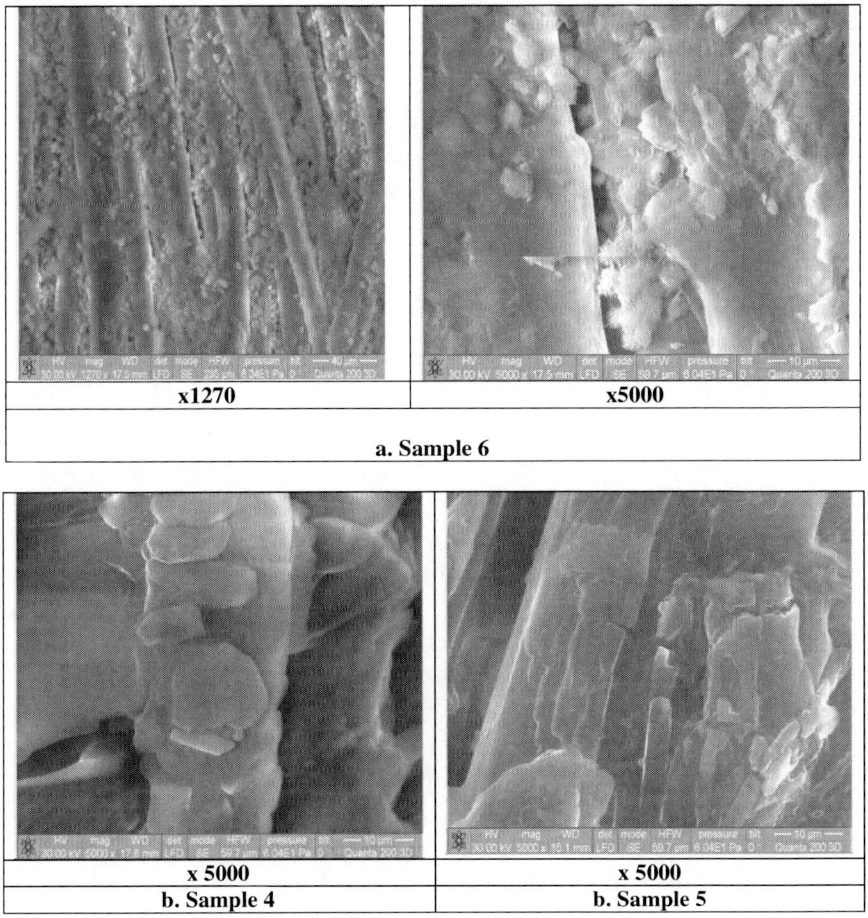

Figure 3. SEM images of some textile composites.

The adhesion strength of ZnO particles on fibrous matrix which depends on the applied surfactant treatment, is different and was tested after repeated washing cycles (1 minute ten times).

(a) (x1200)
sample 3

(b) (x5000)
sample 6

Figure 4. SEM images of some textile composites after repeated washing cycles.

Looking at the SEM images (Figure 4a and b), the adhesion strength of ZnO powder hydrothermally deposited on functionalized polymeric fibrous matrix (linen fibrous support) is superior in case of functionalized surface (Figure 4b) compared with the non-functionalized surface (Figure 4a). Considering the durability of ZnO on the support surface after repeated cycles of washing, the functionalization benefit was evaluated. In case of ZnO powder hydrothermally synthesized without functionalization, after washing the coating particles fell off easily, which might have been caused by the weak attaching force.

According to the images (Figure 4), before treatment the diameters of fibrous supports (individual yarns) are about 10 - 20 μm; after treatment the individual yarns covered by various ZnO aggregates deposition are shown in SEM images. The manner of covering and the shape depend on linen grafting agent assistance. This result is supported by the high number of coordinating functional groups (hydroxyl and glucoside groups) of the MCT-β-CD which can form complexes with divalent metal ions [176]. It might be possible that, during the synthesis time, the majority of the zinc ions were closely associated with the MCT-β-CD molecules. Taking into account the results of previous research, it can be claimed that nucleation and initial crystal growth of ZnO may preferentially occur on MCT-β-CD [210, 119]. Besides, MCT-β-CD, as polysaccharide, showed interesting dynamic supramolecular associations facilitated by inter- and intra-molecular hydrogen bonding, which could act as matrices for nanoparticle growth in size of about 30–40 nm. In a further step, they aggregated to irregular ZnO–CMC nanoparticles (Figures 4a and 4b). SEM images of polymeric fibrous matrices (linen supports) coated with ZnO with assistance of the two surfactants show that the nanoparticles exhibited an approximately lamellar morphology and the particles can be seen to be coated on the fibrous matrix surface (yarn). As result, the fibrous matrices (yarns) surface became coarser after the treatment.

Comparing the influence of the two precursors, in the case of CTAB assistance, on the yarns surface a number of large ZnO particles, covering the yarn as a bark (Figure 4b), can be observed, meaning that in this case the large particles may be formed via precipitation followed by a step-like aggregation process. Due to its high number of coordinating functional groups (hydroxyl and glucoside groups), MCT-β-CD can form complexes with divalent metal ions [210]. It might be possible that the majority of the zinc ions were closely associated with the MCT-β-CD molecules. Moreover, according to the SEM images, the evenness of the polymeric matrix covered with ZnO powder hydrothermally synthesized with assistance of CTAB (Cetyl trimethylammonium bromide) is more adequate than that of ZnO powder hydrothermally synthesized in the presence of Pluronic P123 and possesses good washing fastness. The last one was not been measured, but it was *apriori* evaluated.

This phenomenon can be supported by the fact that the repeated cycles of washing and rinsing did not conduct to the washing away of the ZnO particles; subsequently the zinc oxide proved a low extent of washing fastness. This statement is also in a good correlation with the XRD results, claiming a slight shift of ZnO intensity peaks, meaning that the nucleation of the zinc oxide occurred not only the support surface, but also within the nanocavities, due fibers roughness.

In case of ZnO powder hydrothermally synthesized without any surfactant assistance, the layer particles fell off easily after washing, which might have been caused by the weak attaching force (coordinated bond between ZnO and linen) induced by the deteriorated crystallinity. Moreover, as polysaccharide, MCT-β-CD showed interesting dynamic supramolecular associations facilitated by inter- and intra-molecular hydrogen bonding, which could act as matrices for nanoparticle growth in size of about 30–40 nm. They aggregated to irregular ZnO–CMC nanoparticles in a further step.

II.1.2. Morphological Profile of Hybrid Organic Inorganic Materials Made from *In-Situ* Synthesized Nano-ZnO Particles on Cellulosic Fibrous Matrices (Linen Supports), with the Assistance of MMT and MCT

Looking at SEM photos and comparing the morphology of the two types of samples before and after thermal treatment, some randomly distributed conglomerations are noticeable, in case of *non-functionalized sample*, while on the *functionalized cellulosic fibrous matrix (linen support)*, the particles have bigger dimensions and uniformly distributed.

Some further observations could be made in case of *functionalized cellulosic fibrous matrix*:

- After the thermal treatment at 150°C, the nanoparticles are smaller and partially cover the fiber body (L/MCT-2);
- By increasing the temperature of the treatment up to 450°C, the size of the unfixed ZnO nanoparticles increases and the fibers are uniformly coated (L/MCT-3).
- In case of *non-functionalized polymeric fibrous matrix (linen support):*
- After heating up to 150° C, the nanoparticles sizes randomly augmented (L2);
- A 350°C thermal treatment, induced a diminishment of particles dimension and a relative uniform distribution (L3).

Comparing the morphology of the two studied samples the following discussion should be made. The increasing of the zinc oxide nanoparticles sizes on the functionalized cellulosic fibrous matrix is a consequence of the temperature increasing. In case of non-functionalized cellulosic fibrous matrix, the dimensions of the nanoparticles decreased.

A synergic process occurs – the nuclei increasing and their dissolving, because as a consequence of functionalization, and crystalization nuclei, strongly adherent, whose augmentation is favored by temperature increasing, are noticed. On the other hand, the adherence of particles on non-functionalized matrices is weak, and along with temperature increasing, their destruction takes place.

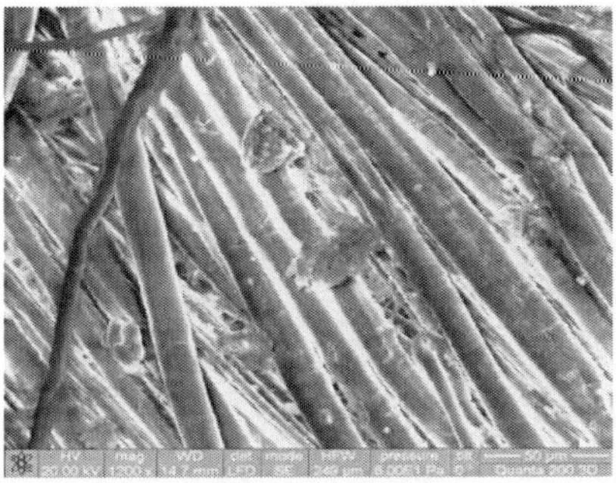

Figure 5. SEM image for L1.

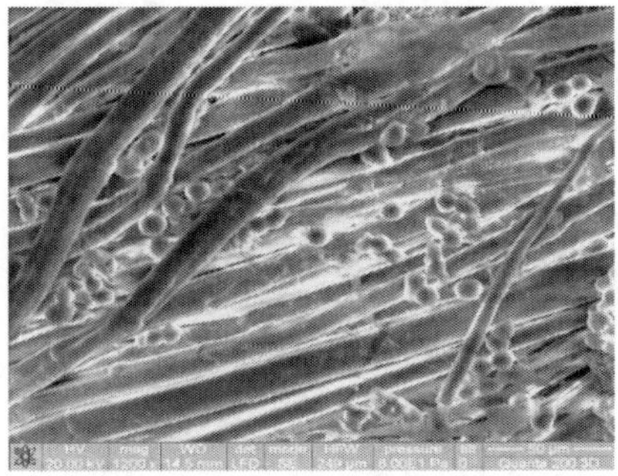

Figure 6. SEM image for L2.

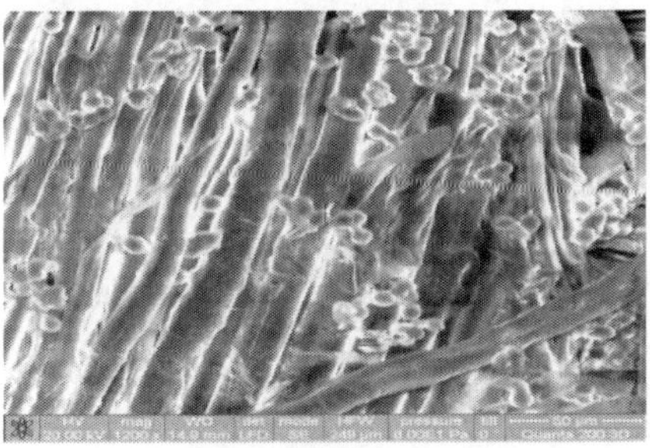

Figure 7. SEM image for L3.

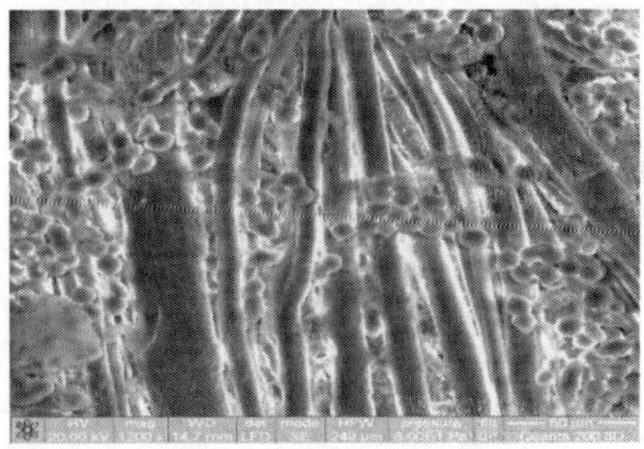

Figure 8. SEM image for L/MCT-1.

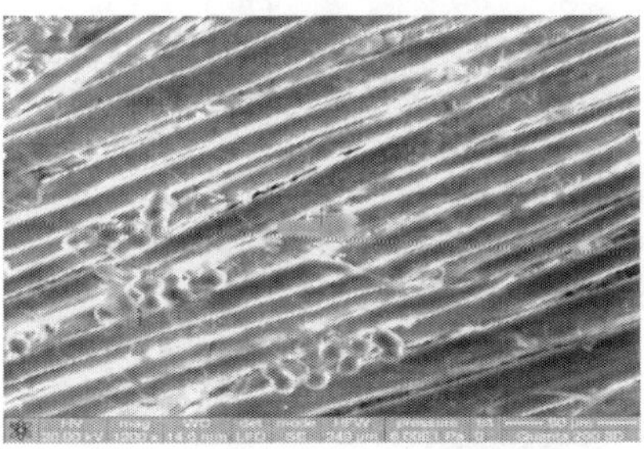

Figure 9. SEM image for L/MCT-2.

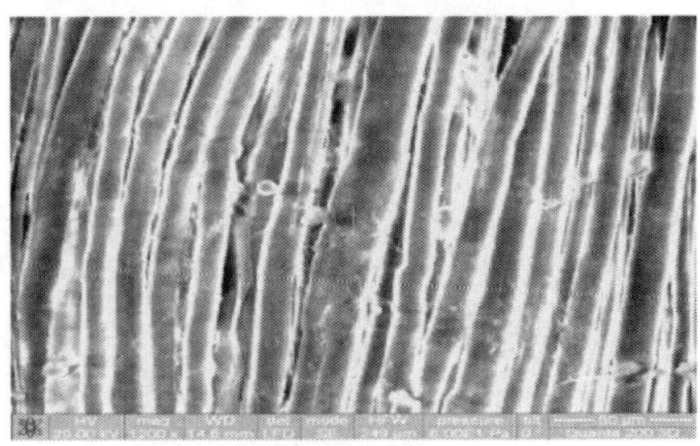

Figure 10. SEM image for L/MCT-3.

II.1.3. Morphological Profile of Nano-Hybrids Made from Nano ZnO and Mono-Chloro-Triazinyl-β-cyclodextrin

SEM images from Figures 11, 12 show a lamellar structure similar to those of biogenic ceramic materials; in particular, calcites formed by the echinoderm intracrystalline and mollusk shell proteins [217]. Randomly distributed particles consisting of nanostructures in MCT matrix, equivalent with a good adhesion between the surface of ZnO nanoparticles and MCT matrix is noticeable.

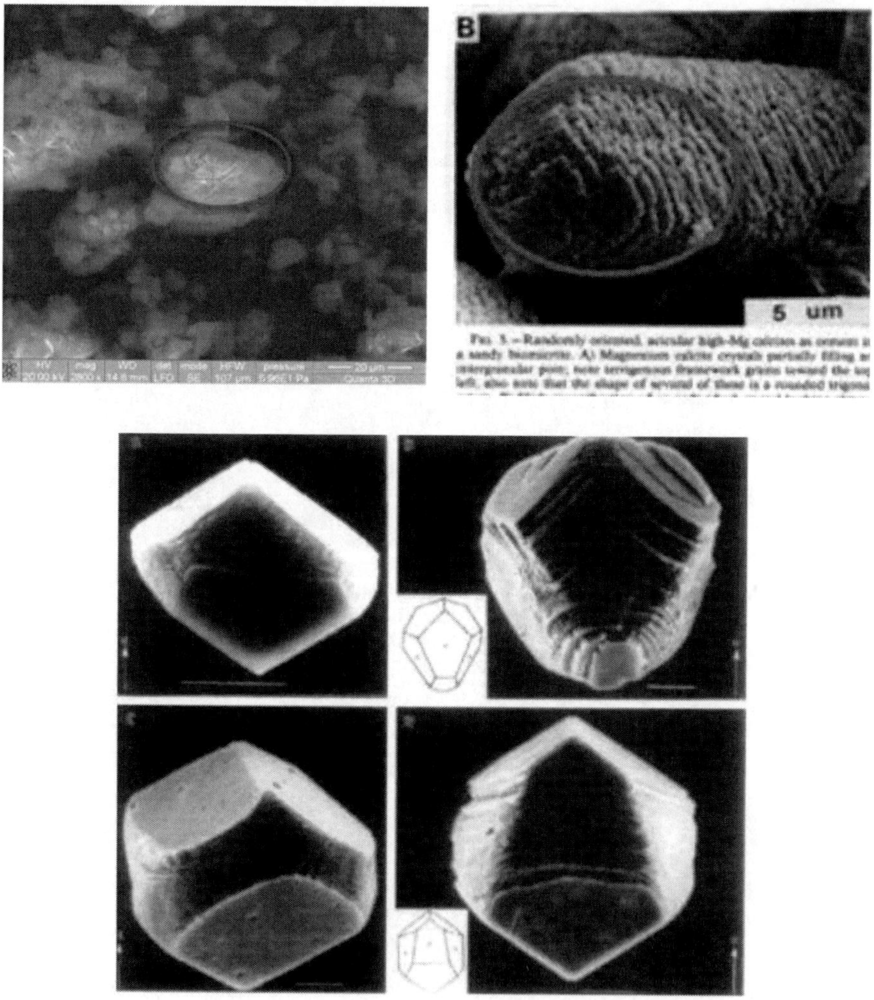

Figure 11. Similarities between calcites from the literature and studied lamellar structures of ZnO nanoparticles entrapped in MCT matrix [217].

Figure 12. Similarities between lamellar structures of ZnO nanoparticles entrapped in MCT matrix at microscale level (2400X, 300 μm) and calcites at macroscopic level (eye observation).

Figure 13. Scanning electron micrograph of zinc oxide embodied in: a. MCT_1 and b. MCT_2 matrices.

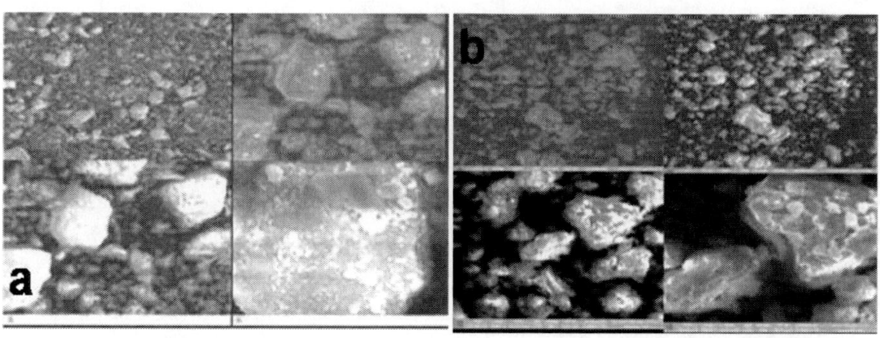

Figure 14. Scanning Electron Micrograph of Zinc Oxide embodied in a - MCT_3 and b - MCT_4 matrices.

The formation of large well-developed rough faces/planes was induced by the ZnO crystals grown in the presence of the MCT. The scale bars represent 20 μm. The new well-developed faces are almost parallel one to each other. It is notable that large, well-defined faces are expressed at relatively high concentrations of MCT (MCT_3 = 1.5%), compared with the untreated material [132].

The formation of micro-pellets with nanometric thickness is induced by the assistance of MCT during ZnO synthesis. This texture is in accordance with the morphology of the single crystal elements, suggesting the influence MCT may also have in the shape of the crystal during growth. Five decades ago, Richard Feynman envisioned these micro-pellets would function as the building blocks for future nanoscale composites [38].

The SEM images (Figures 13, 14) suggest the idea that a two-length-scale hierarchical structure is formed on the surface. A good dispersion of nanoparticles in the Monochlorotriazinyl-β-cyclodextrin nanocavities is noticed; however some aggregated nanoparticles are still visible. The role played by the particles size in determining their adhesion on a hypothetical/potential fibrous matrix is fundamental. This kind of structures is compatible with water repellent properties exhibited by the lotus leaf.

II.1.4. Morphological Profile of Hybrid Organic Inorganic Polymeric Composites Made of PES Fibrous Matrices Covered/Finished with Bulk/Micro Scaled/Powdered ZnO As Candidates to Antimicrobial Character

Figure 15 presents SEM characterization of the prepared samples as presented in Table 3.3. Figure 15a shows a smooth surface of 100% PES fibrous matrix. Small impurities are present. Figure 15b shows PES fibrous matrices modified only with Apretan10% solution. A similar aspect with unmodified matrices can be noticed [143].

Figures 15c and d show the surface of 1% ZnO-PES hybrid composite obtained from water and methanol dispersions. A non-homogeneous coating with ZnO of PES fibrous matrices can be observed.

A rougher coating was obtained by using water dispersion, which is supported by the fact that methanol solvent is to some extent less polar than water, which favors better ZnO coating quality on fibers.

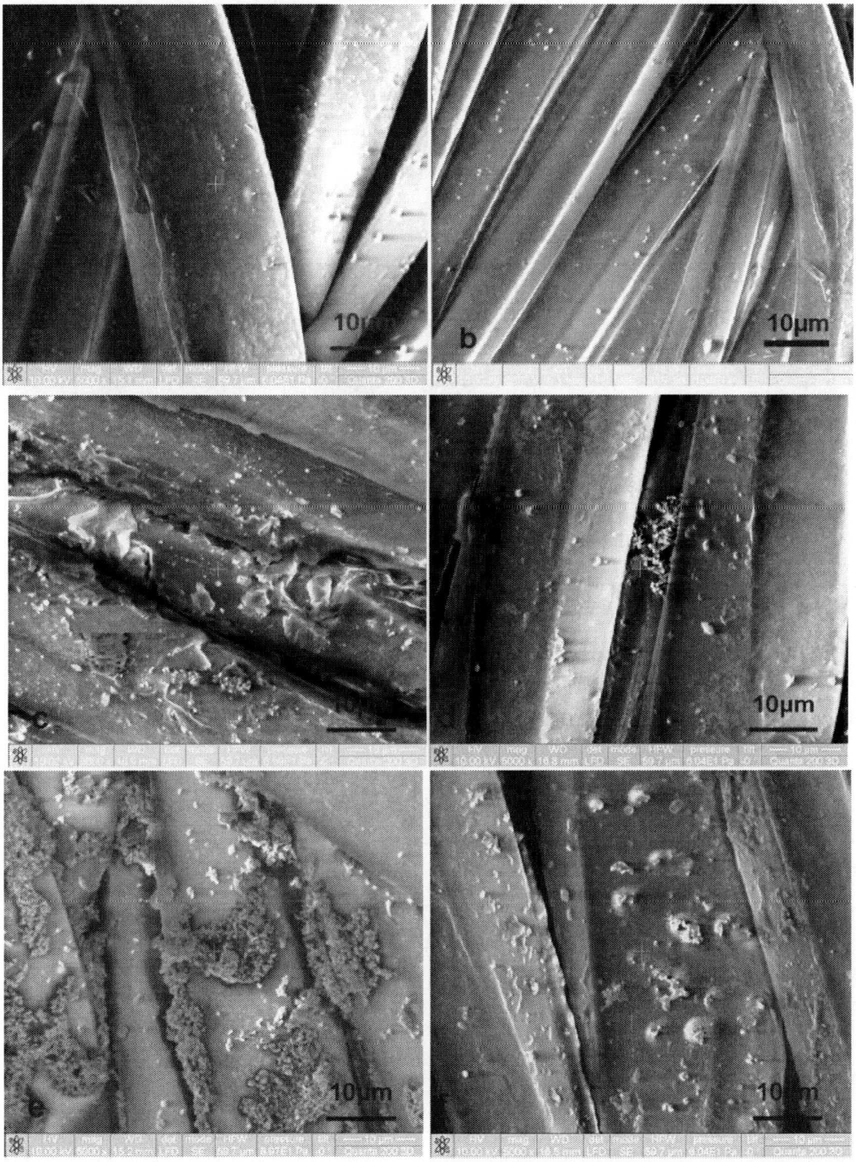

Figure 15. (Continued)

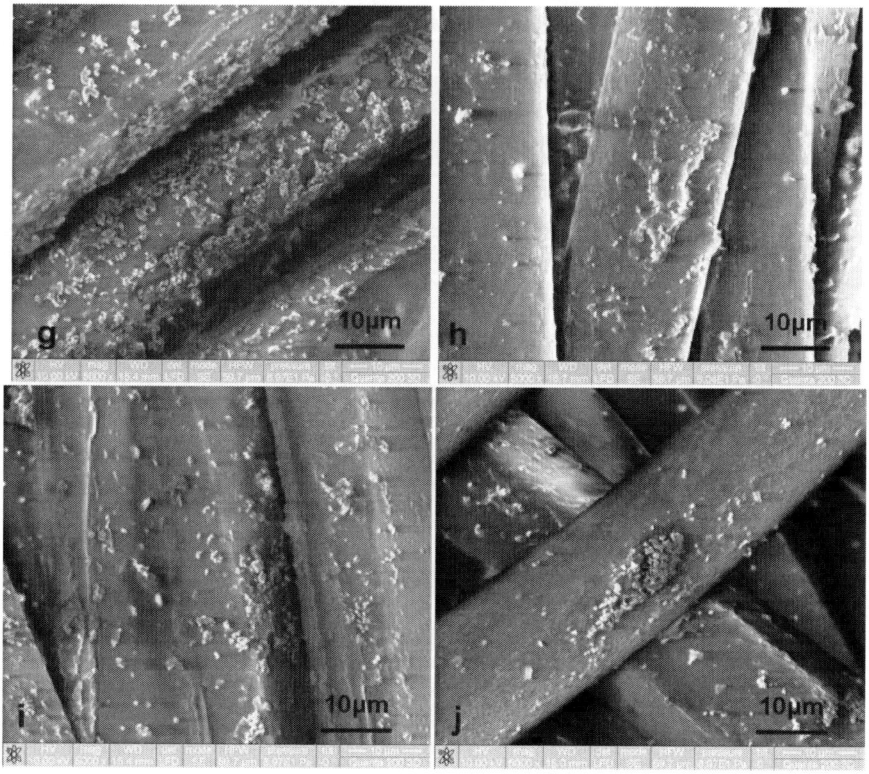

Figure 15. Low vacuum SEM images of ZnO-PES hybrid polymeric composite prepared samples at X5000 magnification and an electron beam energy of 10kV a. 0-PES b.1-PES polymeric matrix acrylic binder c. 2- PES polymeric matrix + 1% ZnO in water d. 2- PES polymeric matrix + 1% ZnO in methanol e. 3-PES polymeric matrix +3% ZnO in water f. 3-PES polymeric matrix +3% ZnO in methanol g. 4-PES polymeric matrix +5% ZnO in water h. 4-PES polymeric matrix +5% ZnO in methanol i. 5-PES polymeric matrix +7% ZnO in water j. 5-PES polymeric matrix +7% ZnO in methanol.

This statement is supported also by different morphology of deposited ZnO than samples coated with ZnO dispersed in water, fact shown by SEM images of the PES fibrous matrices coated with ZnO dispersed in methanol as solvent (Figure 15d, f, h and j). The surface of 3% ZnO-PES hybrid composites from water and methanol dispersions are presented in Figures 15e and f. As observed from these images by increasing the ZnO content leads, as expected, to better coating of PES surface. Further increase up to 5% ZnO leads to a coating morphology as shown in Figures 15g and 15h for ZnO-PES hybrid composite obtained from water and methanol dispersions. Figures 15i and 15j show the surface of 7% ZnO-PES hybrid composite obtained from water and methanol dispersions. A larger concentration lead to agglomerations and a diminished coating quality can be observed. Figure 15f reveals a random distribution of ZnO small size agglomerations, with not dispersed visible particles. Consequently, a higher concentration of finishing agent does not necessary result in an improved efficiency of surface treatment with respect to a better deposition of particles (Figure 15i). By comparing Figure 15 (the lowest concentration of ZnO suspension) and Figure 15i and 15j (the highest concentration of ZnO suspension), an obvious difference of surface coating in terms of aspect and quality can be observed. A better coating of fiber surface than in case of finishing with 3% ZnO suspension was obtained by the treatment with a 5% concentration of ZnO suspension, looking at Figures 15e and g. These results were obtained due to the higher solubility of ZnO in water than methanol. In aqueous medium, ZnO powder nano-particles partially dissolve and subsequently recrystallize on existing ZnO particles in the suspension trough an Ostwald ripening process. Saturation of dissolved ZnO concentration can be achieved faster if the ZnO concentration in the suspension is increased. Growth processes are less obvious and coating homogeneity and uniformity improves. In terms of uniformity level for all concentrations, the covering effect is superior when using methanol as solvent. This analysis suggests that ZnO-PES hybrid composite morphology can be controlled via deposition conditions for tuning of a specific property required for the application. For instance, for a larger surface to volume area, essential in photocatalytic applications, ZnO in water treatment may be a better option than ZnO in methanol. It is obvious that as observable from the respective images corresponding to increasing ZnO concentration for each solvent, larger ZnO concentration in the suspension leads to larger matrix loading.

In terms of r*oughness and topography* of the ZnO coating on PES matrices, the morphological attribute was revealed by AFM method. Figure 16 II.15 present 3D AFM images of surface topography of the 7% ZnO in water and methanol solvents coatings on PES polymeric fibrous matrices. The image for 7% ZnO + methanol PES composite treated for 48 hours was included, as well.

For each concentration and each solvent, matrices local surface roughness and approximate layer thickness were evaluated. For matrices modified with ZnO in water suspension, local roughness was found ranging from few tens of nm to 650 nm; for matrices coated with ZnO in methanol suspension local roughness was slightly lower.

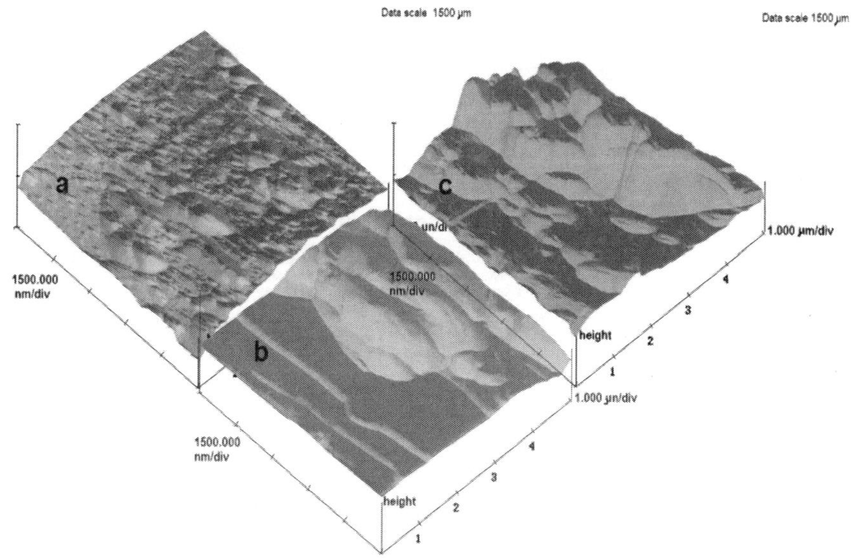

Figure 16. AFM 3D images of: a) PES polymeric matrix / 7% ZnO composite (using water as dispersion medium); b) PES polymeric matrix / 7% ZnO + methanol composite; c) PES polymeric matrix /7% ZnO + methanol composite, for 48 hours.

By comparison samples treated with the same concentration of ZnO suspension, but applied in different conditions it was observed that ZnO in water dispersion corresponded to rougher surface coverage than similar ZnO percentage methanol dispersed. These observations are in accordance with SEM characterization.

In terms of increasing the layer thickness, for the specimens assigned as 7% ZnO powder dispersed in water (Figure 17a), in methanol (Figure 17b) at a treatment duration of 48 h (Figure 17c), one can notice the differences between the cross-sectional profiles.

The average Ra of layer thickness is 630-650 nm, by estimating the cross section analysis, for the specific samples. The variation of thickness layer was found as follows: from about 1μm up to about 7 μm with increasing the ZnO concentration in water and from about 700 nm up to about 6 μm for ZnO in methanol. In case of passive ZnO in methanol treatment, a duration of 48 hours leads to much lower thickness than direct impregnation case for similar concentrations.

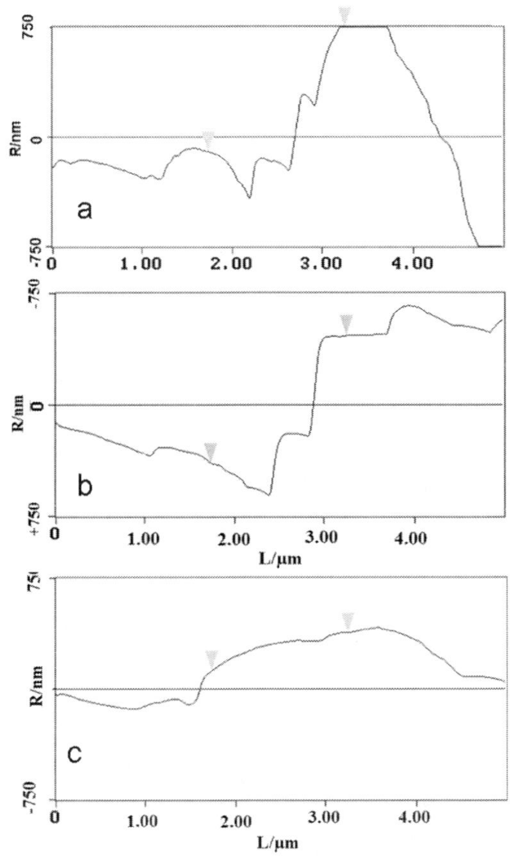

Figure 17. Cross-section analysis of: a) PES polymeric matrix / 7% ZnO composite (using water as dispersion medium); b) PES polymeric matrix / 7% ZnO + methanol composite; c) PES polymeric matrix /7% ZnO + methanol composite, for 48 hours.

II.1.5. Morphological Profile of Hybrid Organic Inorganic Polymeric Composites Made from PES Fibrous Matrices Covered/Finished with Bulk/Micro Scaled/Powdered ZnO As Candidates to Anti-UV and High Hydrophobic Character

SEM microphotographs of polyester matrices coated with ZnO are shown in Figure 18. The following observations should be made. The surface of uncoated polyester fabric is smooth (Figure 18a), showing only some small isolated particles, probably impurities. By an impregnation with ZnO solution, a cover of matrices with a layer, showing different morphology with increasing ZnO concentration of the finishing agent was obtained. A thin layer, with lamellar morphology and good uniformity, develops on the 1% ZnO sample (Figure 18b). By augmenting the ZnO concentration, this layer increases in thickness, showing uniform coating of the fibers for 3–5% ZnO (Figure 18c). Nevertheless, for higher concentration of the finishing agent, the dispersion of the microparticles is poorer, and the tendency to agglomerate visibly conducts to non-uniform coating of the polymeric matrices (Figure 18d).

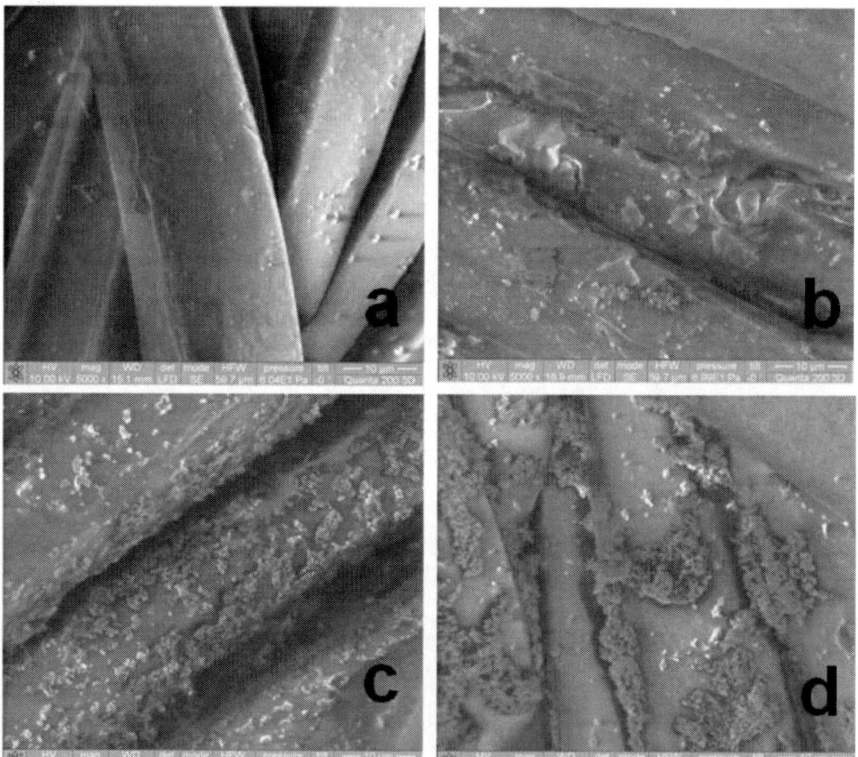

Figure 18. SEM microphotographs of polyester matrices coated with ZnO: a) uncoated, b) 1% ZnO, b) 5% ZnO, d) 7% ZnO.

Therefore, at least in the case of microparticle powder, satisfactory dispersion of the particles, ensuring uniform coating of fibers, corresponds to ZnO concentration lower than 5%.

II.2. Summary of EDX Data on Polymeric Fibrous Matrices Covered with Bulk/Micro and Nano ZnO

II.2.1. Elemental Analysis for Composites Made from *In-Situ* Synthesized Nano-ZnO Particles on Cellulosic Fibrous Matrices (Linen Supports), with the Assistance of CTAB and P123

In terms of elemental analysis the results of the EDX measurements for sample 6 illustrated in Figure 19 and Table 3.1, show that surface composition contains approximately 47% ZnO, meaning that ZnO phase represented almost half of the sample mass.

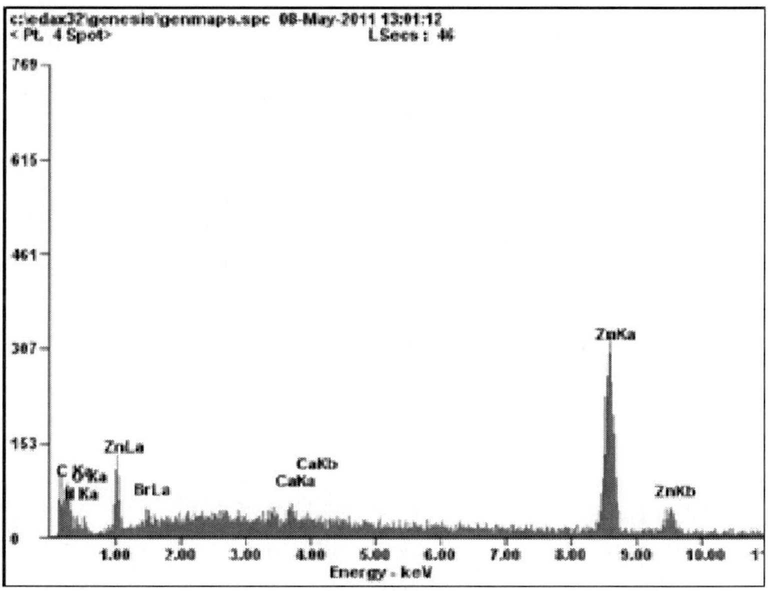

Figure 19. EDX analysis (sample 6 – from Table II.1) (Wt: weight percent, At: atomic percent).

Table II.5. Surface composition from EDX measurements at sample 6 – from Table 3.1.

Element	*Wt%*	*At%*
CK	25.34	46.47
NK	15.45	24.30
OK	08.91	12.27
BrL	01.90	00.52
CaK	00.63	00.35
ZnK	47.77	16.10
Matrix	Correction	ZAF

II.2.2. Elemental Composition of Composites Made from Nano-ZnO Particles Applied on Cellulosic Fibrous Matrices (Linen Supports) with the Assistance of MMT and MCT

Figures 20-25 and Tables II.6 and II.7, indicate that ZnO nanocomposites contain different percentage content of zinc oxide.

For non-functionalized cellulosic fibrous matrices (set linen supports):

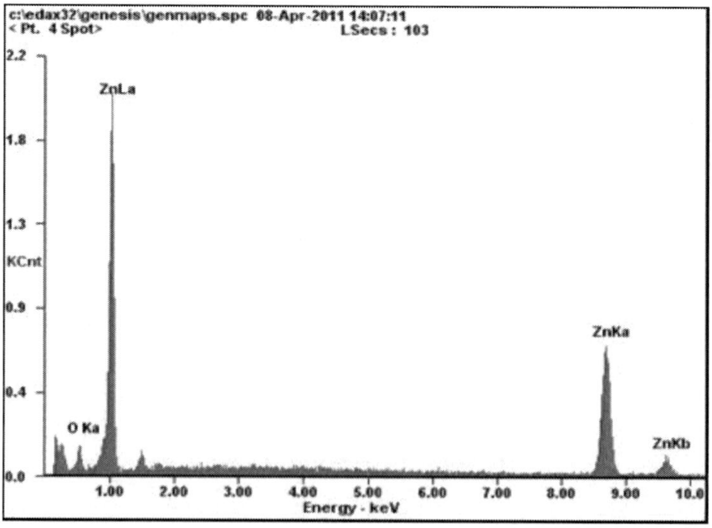

Figure 20. EDX analysis for L1.

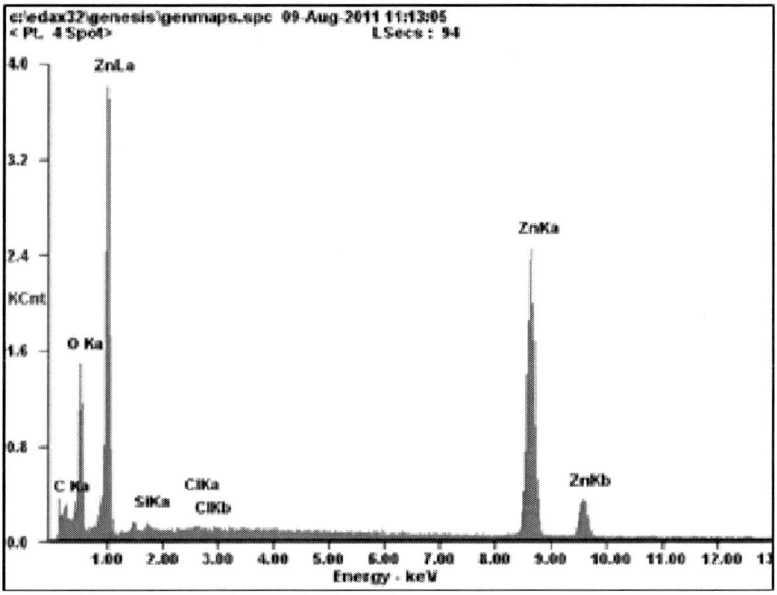

Figure 21. EDX analysis for L2

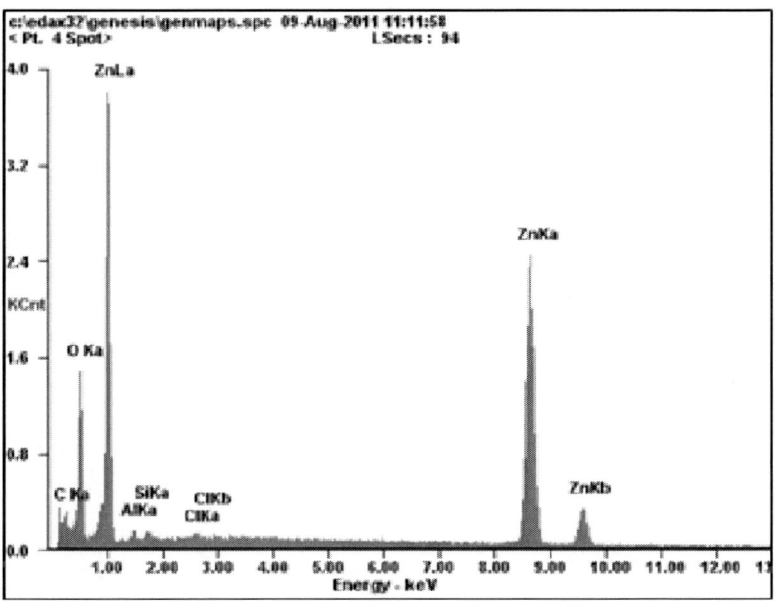

Figure 22. EDX analysis for L3.

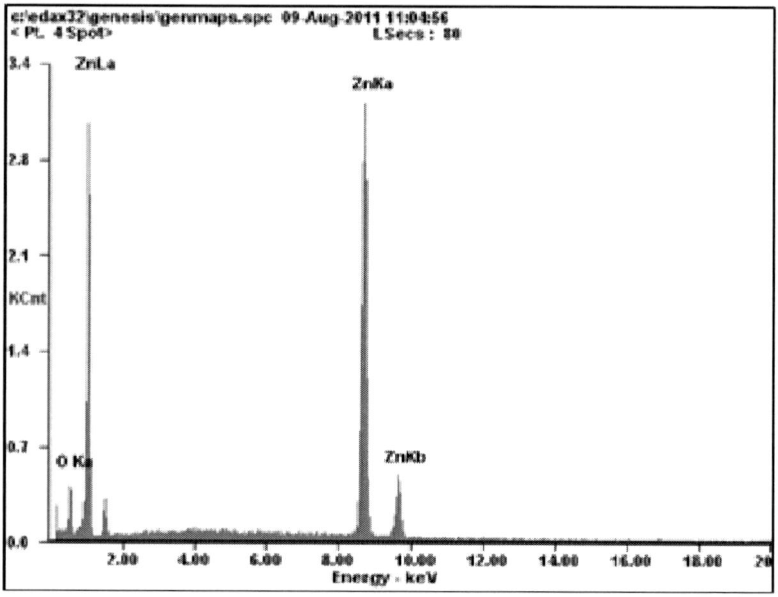

Figure 23. EDX analysis for L/MCT-1.

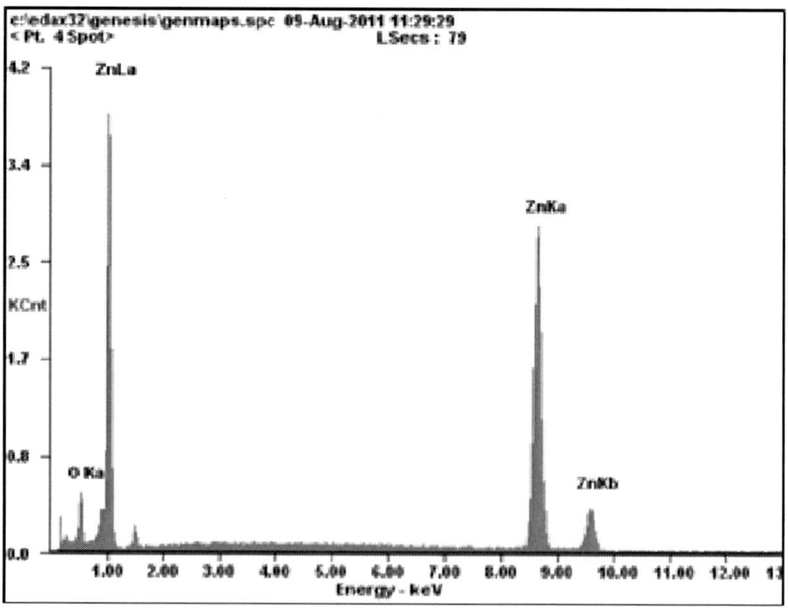

Figure 24. EDX analysis for L/MCT-2.

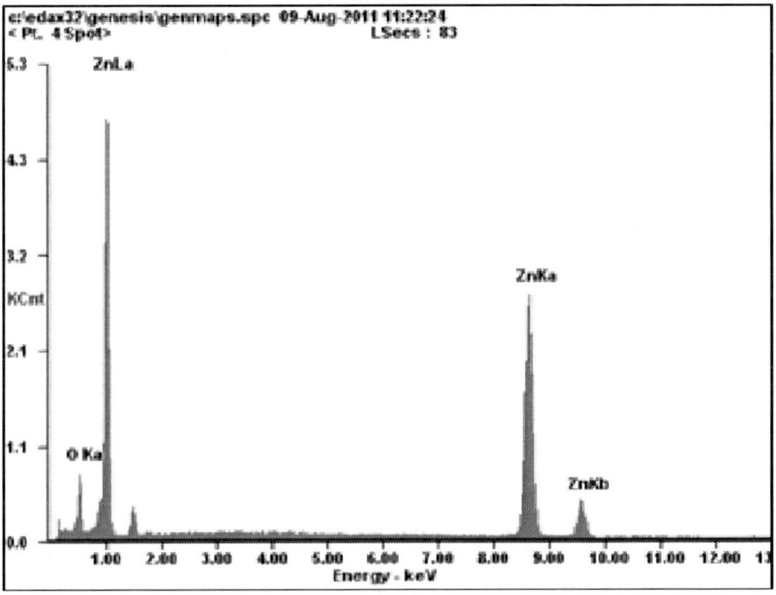

Figure 25. EDX analysis for L/MCT-3.

Table II.6. Surface composition from EDX measurements /Wt: weight percent, At: atomic percent, for sample L1, L2, L3

Element		*Wt /%*	*At /%*
OK	L1	28.91	62.43
	L2	27.40	42.74
	L3	27.08	42.18
ZnK	L1	71.09	37.57
	L2	54.43	20.78
	L3	53.64	20.45
CK	L1	17.14	35.61
	L2		
	L3	17.06	35.40
SiK	L1	00.77	00.68
	L2		
	L3	00.76	00.67
ClK	L1		
	L2		
	L3	00.26	00.18
AlK	L3	01.2	01.11

Table II.7. Surface composition from EDX measurements/Wt: weight percent, At: atomic percent, for L/MCT-1, L/MCT-2, L/MCT-3

Element		*Wt /%*	*At /%*
OK	L/MCT-1	08.94	28.6
	L/MCT-2	12.05	35.89
	L/MCT-3	15.35	42.56
ZnK	L/MCT-1	91.06	71.37
	L/MCT-2	87.95	64.11
	L/MCT-3	84.65	57.44

As a conclusion, a decreasing of zinc oxide content with the increasing of the treatment temperature is remarkable (Table II.6 and II.7).

II.2.3. Elemental Composition of Nano-Hybrids Made from Nano ZnO and Monochlorotriazinyl-β-cyclodextrin

Figure 26 indicates that the hybrids contain approximately 80% ZnO.

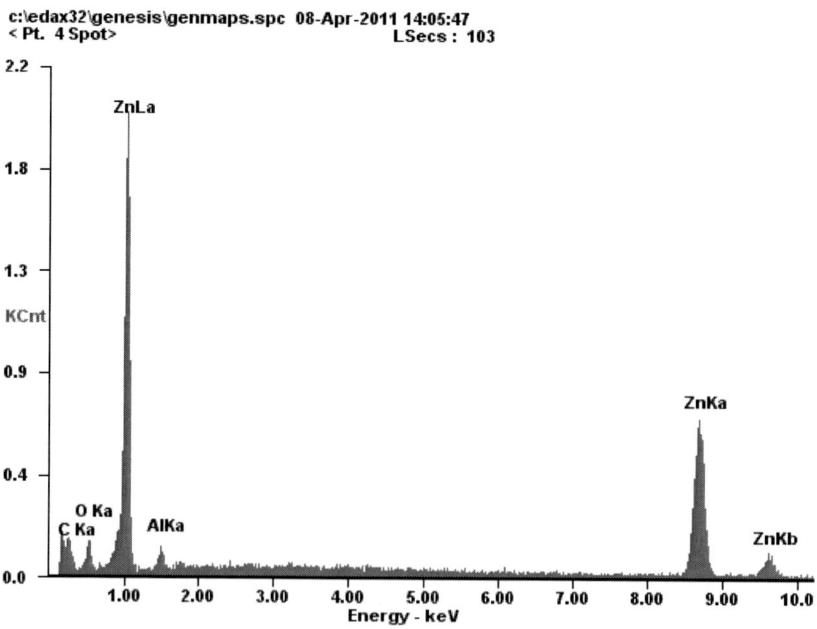

Element	*Wt%*	*At%*
CK	14.23	43.43
OK	03.93	09.01
AlK	02.08	02.83
ZnK	79.76	44.74

Figure 26. EDX analysis (Wt: weight percent, At: atomic percent).

II.2.4. Elemental Composition for Hybrid Organic Inorganic Polymeric Composites Made of PES Fibrous Matrices Covered/Finished with Bulk/Micro Scaled/Powdered ZnO As Candidates to Antimicrobial Character

Because high energy electron beam quickly damages the fibrous matrices, this analysis has only qualitative value. In Figure 27, a typical EDX spectrum (for sample prepared from 3% ZnO in methanol dispersion with the highest Zn content) is presented, the inset presenting also the compositional analysis in mass and atomic percents. Since the material presents some limitations in the specific analysis conditions. These are approximate values.

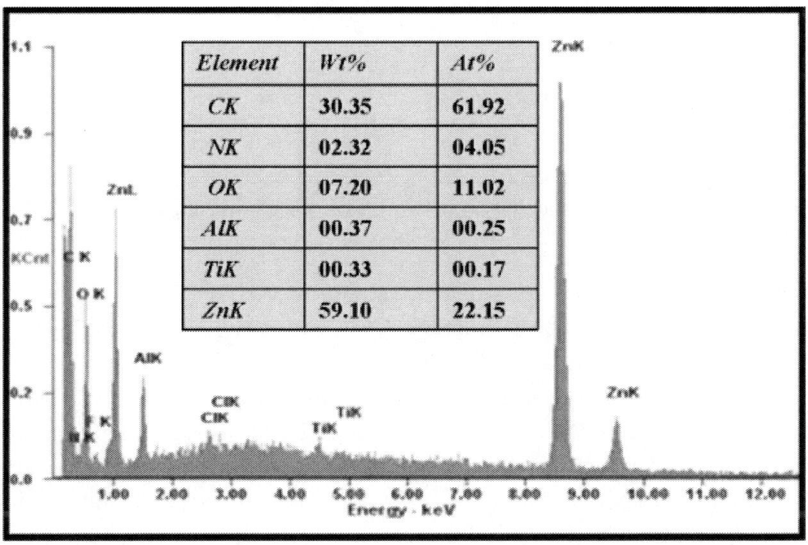

Element	*Wt%*	*At%*
CK	**30.35**	**61.92**
NK	**02.32**	**04.05**
OK	**07.20**	**11.02**
AlK	**00.37**	**00.25**
TiK	**00.33**	**00.17**
ZnK	**59.10**	**22.15**

Figure 27. EDX spectrum for sample prepared from 3% ZnO in methanol dispersion with the highest Zn content.

II.3. XRD Data on Normalized Intensities (%) of ZnO Architectures. XRD Patterns Interpretation

II.3.1. XRD Data Of Hybrid Organic Inorganic Materials Made From In-Situ Synthesized Nano-ZnO Particles on Cellulosic Fibrous Matrices (Linen Supports), with the Assistance of CTAB and P123

Figure 28 presents X-ray diffraction patterns of samples 4-7 compared with reference 2:

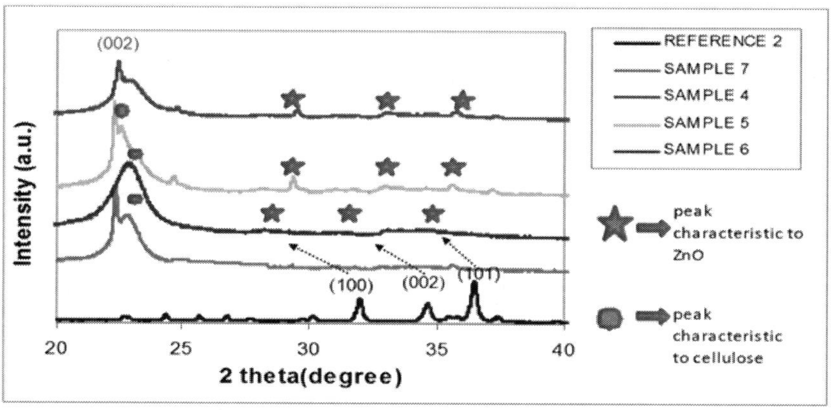

Figure 28. (Color online) XRD patterns of: Reference 2; Sample 4; Sample 5; Sample 6; Sample 7; the arrows indicate the peaks shift, in terms of working conditions. The height of the peaks was multiplied by a 40 factor.

Figure 28 reveals selected-area diffraction patterns (2θ=20-40°) of the obtained samples. All the diffraction lines are assigned to the wurtzite hexagonal phase structure, according to the literature [119].

The patterns belonging to hybrid composites (sample 3-6) reveal both the presence of the peaks positions that matched well with those of the ZnO XRD pattern - lines (100), (002) and (101) - and the main peak of cellulose - linen (002) [203]. A high content of deposited ZnO revealed by small relative intensity of the peaks of the ZnO–linen composites could be considered as "the dark side" of the research, since it is not well correlated with the EDX analysis. The ZnO diffraction lines shift (samples 4 and 5) denotes the fact that the growth of the ZnO takes place not only on the matrix surface, but also inside the nanocavities due to the fibers roughness. By analyzing Figure 28, the intensities of the diffraction peaks decrease when the synthesis takes place with the assistance of the surfactant, which means avoiding crystal growth, in these working conditions.

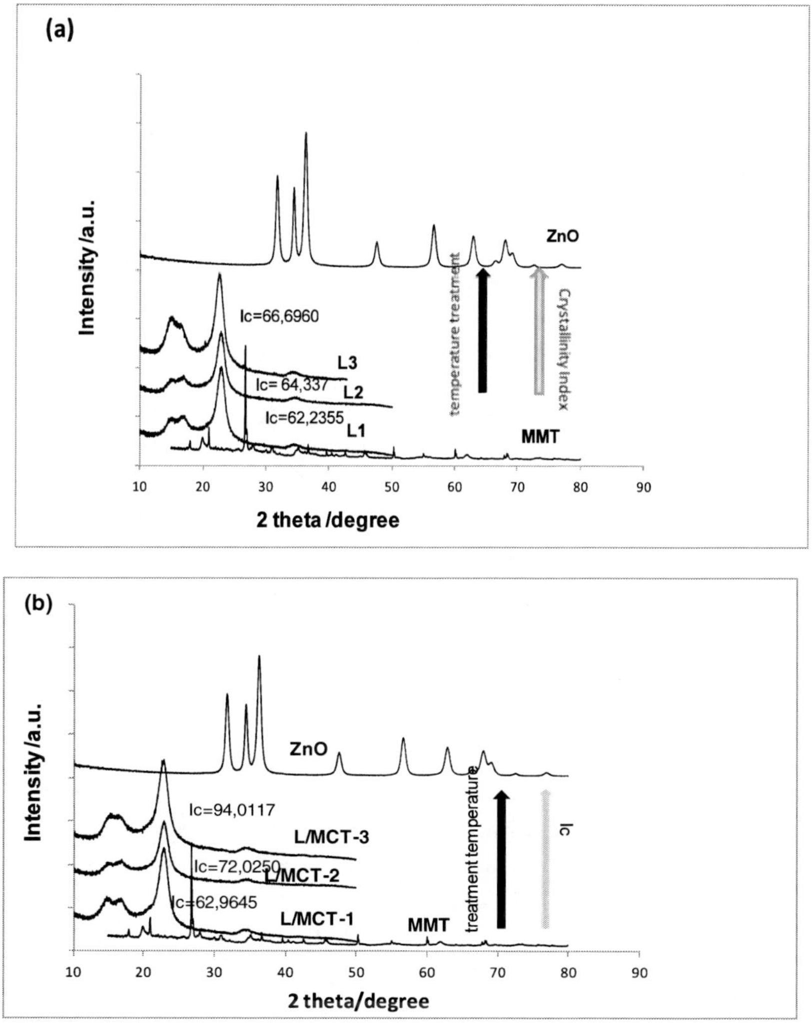

Figure 29. XRD patterns of: a) MMT, ZnO, L1, L2, L3 and b) L/MCT-1, L/MCT-2, L/MCT-3.

II.3.2. XRD Data of Hybrid Organic Inorganic Materials Made from *In-Situ* Synthesized nano-ZnO Particles on Cellulosic Fibrous Matrices (Linen Supports), with the Assistance of MMT and MCT

In case of specimens associated with the thermal response, the XRD patterns below exhibited, for specific peaks of flax fibers, meaning: four well-defined peaks at 15.1°, 16.8°, 22.0° and 34.4°, the values of 15.1° and 16.8° for the 2 θ reflection, corresponding to the 110 crystallographic planes, respectively (Figure 29).

The other two peaks at 22.0 and 34.4 correspond to the 002 and 004 planes, which is in accordance with the literature [186].

Due to the large amount of amorphous regions present in cellulose, and also to the presence of amorphous lignin and hemicelluloses, in case of the functionalized polymeric fibrous matrix, three peaks appeared around 15.5°, 22.0° and 34.4°, corresponding to the 110, 002 and 004 planes, respectively (Figure 29b), which agrees with the results of Tserki [178].

Due to the small concentration of ZnO solution, in our opinion, an overlapping of the peak characteristic to flax fibers (Pos.[°2Th.] ranging between 30-40 degrees) to those specific to both ZnO and MMT.

For L2, it is noticeable that the apex of the peak specific for MMT –is also found in the L3, meaning that MMT was not thermally degraded. Moreover, the crystallinity index increased, from a thermal stage to other. Generally, through calcination process, it can be stated that the CI is maintained approximately constant, meaning that the assistance of MMT prior to the application by padding of zinc oxide, stabilizes in a certain extent the fibrous polymeric matrix.

Referring to the functionalized type and probably due the existence of MCT on the fibrous matrix surface, playing the role of entrapping/entrapment of the ZnO nanoparticles, as it can be found in a previous study, a more enhanced CI was obtained. However, the XRD analysis revealed the fact that the entrapment was not good enough. This can be explained also by the fact that the repeated cycles of washing and rinsing conducted to the removal of the byproducts, contributing also to the washing away of the ZnO unbound particles.

According to previous studies, a correlation between crystallite size and CI should be done, in order to generate a clear overview regarding the crystallinity of the studied supports and their novel nanoscale pattern surface achieved by the last one, responsible for the thermal protection/barrier [119, 75].

For the two different specimens, the d spacing of the (020) planes is listed in Table II.8. The values of d for 3–non-functionalized polymeric fibrous matrix with ZnO thermally non-treated and 4–functionalized polymeric fibrous matrix with ZnO treated at 150°C were almost equal. However, the spacing becomes larger for the 3–non-functionalized polymeric fibrous matrix with ZnO treated at 150°C and still larger for 4–functionalized polymeric fibrous matrix with ZnO thermally non-treated. Changes in the crystallinity index and in the size of the crystals perpendicular to the (020) plane follow a reversed tendency. Spacings of more than 0.4 nm for the (020) planes as shown in Table II.8, correspond generally to cellulose micro fibrils of low crystallinity and small lateral size, which contrasts with the high crystallinity and large size of the cellulose micro fibrils contained in the cell wall. The least crystalline specimens are obviously more distorted, which results in a looser packing of the chains and hence larger unit cell dimensions. After conversion to cellulose I1 by mercerization, crystallinity index and crystal size tend toward similar values for all the celluloses as shown in Table 4.28. The crystallinity index for the L1, L2, L3 specimens remains in a narrow range of 0.622-0.669 while it varies from 0.6296 to 0.9401 for the functionalized polymeric fibrous matrices. The crystal size severely decreases for L/MCT-1, L/MCT-2, L/MCT-3 at the first stage of thermal treatment from 26.74 nm to 10.96 nm, while it varies from 10.08 nm to 10.62 nm in non-functionalized linen support through the treatment, with an increase in temperature. An evident convergence of the crystallinity and of the crystal size during the thermal treatment occurs. The results indicate clearly that in non-functionalized sample of low crystallinity, the summative effect of ZnO, MCT occurrence and the progressive augmentation of the temperature induced a slow increasing of crystallinity and crystal size, while for functionalized specimen, a synergic result occurred: the crystallinity index remarkably increased, and also the crystallite size. These trends would not be expected without the influence of the chemicals assistance: MCT as functionalization agent, on one hand, and ZnO as stabilizer, on the other hand.

The results are more in line with the idea that the efficiency of ZnO particles coating on previously functionalized polymeric fibrous matrices involves an obvious enhanced crystallinity. On the other hand, the non-functionalized specimen shows a slight improvement in crystallinity, due the existence of the ZnO nanoparticles, but with the absence of the agent of entrapment, the crystallinity is not complete.

This concept is compatible with the data interpretation provided by FT-IR analysis. There is no any definitive proof that the picture presented above is correct. However, as far as the authors are aware, the present study is the first report that larger crystals and higher crystallinity may be obtained by the synergic assistance of MCT as functionalization agent and ZnO, as thermal barrier-conferring compound.

From the Table II.8 some observations should be stated:

- At L/MCT-2, L/MCT-2, L/MCT-3 specimens, CI increases with the decreasing of the crystallites nanoparticles;
- In case of L1, L2, L3, there is a proportionally increasing of CI with the augmentation of nanoparticles size.

Table II.8. Correlation between crystallite size and CI

Sample	Crystallite size perpendicular to the (020) planes, by XRD measurement/nm	d spacing of the (020) planes distance/Å	C.I./crystallinity index
L1	10.08	0.3889	0.622
L2	10.32	0.8786	0.643
L3	10.62	0.3928	0.669
L/MCT-1	26.74	0.9060	0.6296
L/MCT-2	11.92	0.387	0.7202
L/MCT-3	10.96	0.3929	0.9401

Some Weak Points of the Research

The calculation of CI was made from the ratio of the height of the 002 peak (I_{020}) and the height of the minimum (IAM) between the 020 and the 101 peaks, as shown in Figure 29 [42]. This method is useful for comparing the relative differences between samples; however, we suggest that it should not be used as a method for estimating the amount of crystalline and amorphous material in a cellulose sample for the following reasons. The minimum position between the 002 and the 101 peaks (I_{AM} which is at about 18.3° in Figure 29a) is not aligned with the maximum height of the amorphous peak. The apex of the peak that is due to amorphous cellulose is likely to be higher than 18,3°. From the peak deconvolution method, the amorphous peak was predicted to be at around 21.5°. Thus, the I_{AM} value for the height method is significantly underestimated, resulting in an overestimation of the CI.

Although there are at least four crystalline peaks, only the highest peak (002) is used in the calculation. This fact excludes contributions from the other crystalline peaks, putting too much emphasis on the contribution from one alignment of the cellulose crystal lattice. Peaks in the cellulose diffraction spectrum are very broad and vary considerably in their width. A simple height comparison cannot be expected to provide a reasonable estimate of cellulose crystallinity, as it neglects variation in peak width, which can also be affected by crystallite size [112].

The relative height to the minimum can only be taken as a rough approximation of the contribution of amorphous cellulose to the cellulose diffraction spectrum.

If we compare the two sets of probes, the functionalized group has CI higher, meaning an improved crystallinity, while the non-functionalized one, even though it also revealed an augmentation of the crystallinity, it was a more reduced one. The perspective research will be oriented to more improved technique of fixing the ZnO nanoparticles in order to achieve/add multifunctionality of the studied fibrous nanocomposites.

II.3.3. XRD Data of Nano-Hybrids Made from Nano ZnO and Monochlorotriazinyl-β-cyclodextrin

XRD patterns of ZnO nanocrystals entrapped into MCT_1, MCT_2, MCT_3 and MCT_4 polymer matrices are presented in Figure 30.

The peaks assigned to diffractions from various planes correspond to hexagonal close packed structure of zinc oxide. The broadening of peaks was observed mainly due to the nano-size effect. The broad reflection at 25° is due to the low crystallinity of the soluble MCT. Presence of MCT in the completely washed nano-ZnO indicates their strong binding nature. Peaks at scattering angles (2θ) of 26.5737, 33.72, 38.03, 51.87, 64.71 and 71.32 indicate the crystallinity of the synthesized solid.

The XRD pattern is identical to the hexagonal phase with Wurtzite structure (hexagonal phase) unit cell parameters a = b = 3.248 Å and c = 5.2 Å.

Rather broad diffraction maxima indicate very small size of the crystallites similar to the size of subunits observed by the SEM. Meanwhile, no diffraction peaks from other species could be detected, which indicates that all the precursors were completely decomposed during the sol-gel process. It is noteworthy that encaving/entrapment process of ZnO particles into cyclodextrin matrices induced the augmentation of compound crystallinity with respect to multiplying of peaks (64.71 and 71.32) for ZnO-MCT_2, ZnO-MCT_3 and ZnO-MCT_4 indexed samples. In our case, the increasing of MCT concentration did not mean an improvement of crystallinity, since the XRD pattern attributed to ZnO-MCT3 and ZnO-MCT4 compounds seem to have many similarities.

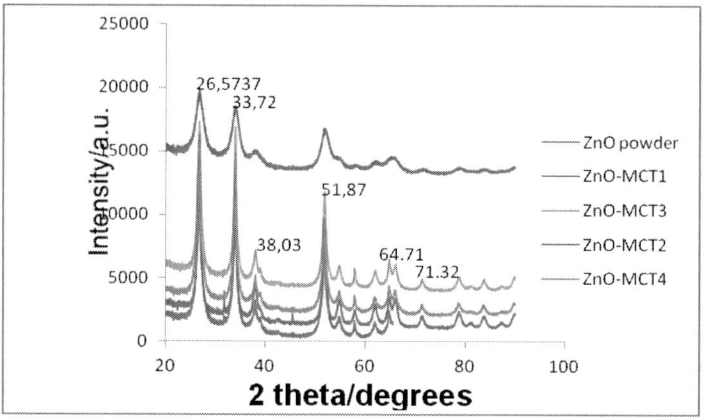

Figure 30. XRD pattern of nano ZnO entrapped in different concentration of MCT (Monochlorotriazinyl-β-cyclodextrin); a - ZnO-MCT_1; b - ZnO-MCT_2;c - ZnO-MCT_3; d- ZnO-MCT_4.

II.4. Specific Surface Area Data Summary of Hybrid Organic Inorganic Materials Made from In-Situ Synthesized Nano-ZnO Particles on Cellulosic Fibrous Matrices (Linen Supports), with the Assistance of CTAB and P123. Water Vapors Sorption Responsivity

Isothermal studies can be performed as a function of humidity (0-95%) in the temperature range 5° C to 85° C, with an accuracy of ± 1% for 0 - 90% RH and ± 2% for 90 - 95% RH. The relative humidity (RH) is controlled by wet and dry nitrogen flows around the sample. The RH is held constant until equilibrium or until a given time is exceeded, before changing the RH to the next level.

In the sample room the vapors pressure was achieved by 10 steps of 10% humidity, each of them having a time of equilibrium setting between 10-20 minutes. In each phase, when the equilibrium is reached the weight adsorbed by the sample is measured by electromagnetic compensation between tare and sample. The apparatus has an anti-condensation system for the cases vapors pressure is very close/near to that of saturation. The cycle is finished by decreasing in steps vapors pressure, in order to obtain desorption isotherms, as well.

The drying of the samples is performed in nitrogen flow (250 mL/min) at 25°C, until the sample weight reached a constant value, at a relative humidity less than 1%, prior to measuring of sorption-desorption isotherms.

The sorption/desorption isotherms recorded are shown in Figure 31.

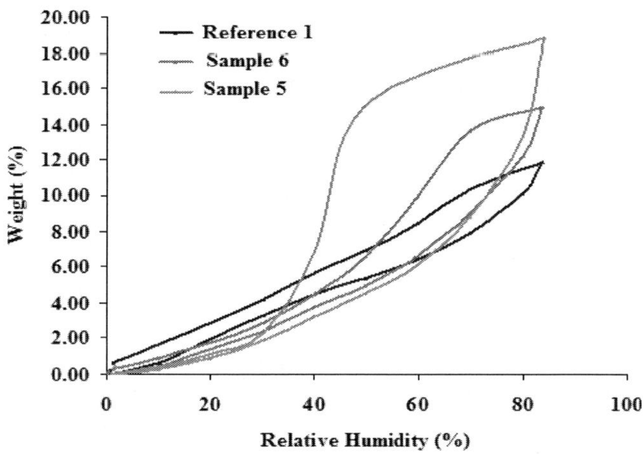

Figure 31. Comparative plots of rapid isotherms for water vapors sorption for the studied samples: Reference [1]; Sample 5; Sample 6.

Looking at plots, the reference sample (the cellulosic fibrous matrix – yarn - unfunctionalized) has a smaller sorption capacity compared to that of Sample 6 and Sample 5. High values of water vapors sorption capacity for the two last samples prove the fact that the material surface becomes more hydrophilic, more porous, respectively as it could be observed from hysteresis shape.

One of the main objectives of book was highlighting the adsorptive attributes, by improving of ZnO synthesis conditions. Consequently, the role of P123 in the ZnO synthesis was to obtain a composite with a higher porosity, in order to achieve the surface hydrophilicity, since there is direct correlation between porosity and hydrophilicity [144].

For those two compounds the shape of the moisture sorption isotherms is similar to those characteristic of mesoporous materials (type IV, according to IUPAC classification – with low sorption at low water vapor sorption (adsorption/desorption), moderate sorption at average humidity and rapidly increasing water sorption at high humidity). This type of isotherm describes the sorption behavior of hydrophilic material [22].

By exposing a material to environmental water vapors, the water molecules firstly react with surface polar groups and form a molecular monolayer.

Based on the sorption studies, the IGAsorp software allows an evaluation of both monolayer and surface area value, by using BET (Brunauer-Emmett-Teller) model (Table II.9).

Table II.9. The main parameters of (water vapors) sorption-desorption isotherms for the studied samples

Sample	Sorption capacity (%d.b.)	BET analysis	
		A_{BET} (m^2/g)	Monolayer (g/g)
Reference 1	11.89	157.010	0.044
Sample 6	14.93	213.99	0.060
Sample 5	18.89	321.39	0.091

For modeling of the sorption isotherms BET (1) equation is very often used:

$$W = \frac{W_m \cdot C \cdot RH}{(1 - RH) \cdot (1 - RH + C \cdot RH)} \quad (1)$$

where:

W- the weight of adsorbed water, Wm- the weight of water forming a monolayer, C – the sorption constant, *p/po=RH*- the relative humidity.

Sorption isotherms up to a relative humidity of 40% are described by BET model and they are related to the sorption isotherm and material type. This method is mainly limited for isotherms of IInd type but can describe the isotherms of I, III and IV type [52, 15]. The increasing water sorption is reflected both by the augmentation of monolayer and surface area values calculated with BET model (Table II.9).

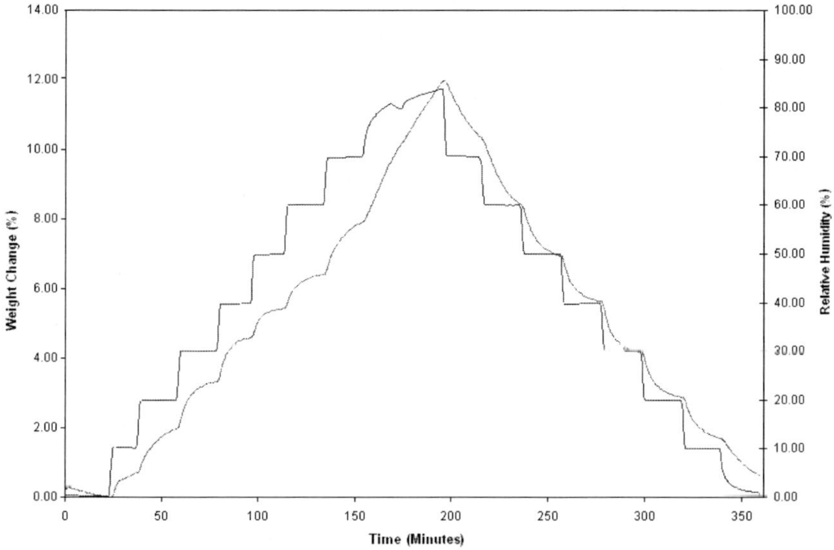

Reference 1

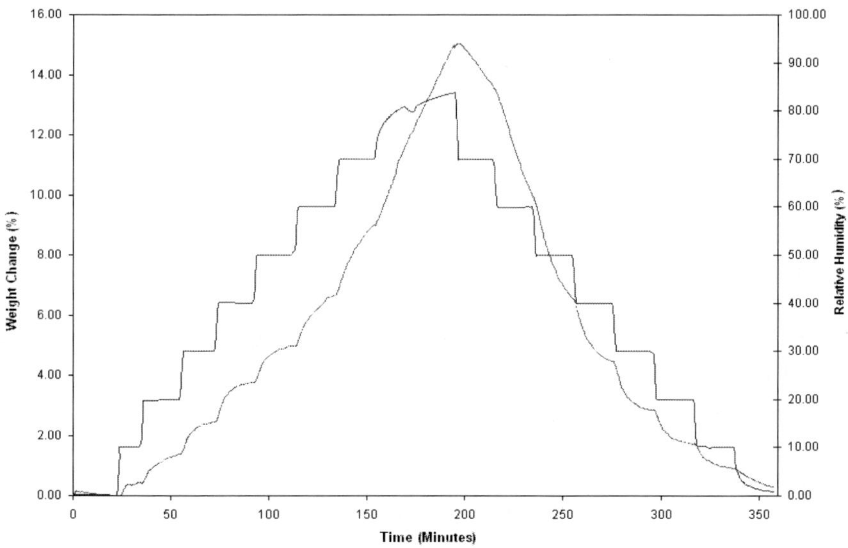

Sample 6

Figure 32. Kinetic curves for sorption/desorption processes of water vapors in the studied samples: Reference [1]; Sample 6.

The kinetic curves for humidity (water vapors) sorption/desorption processes for two of the samples are displayed in Figure 32. It is noticed that it takes more time for equilibrium setting for sorption processes than for desorption. Sorption rate is smaller than that of desorption.

After the samples were kept at RH=90%, until the mass became constant, in Table II.10 the dynamic moisture sorption capacity was calculated using the equation written below:

$$\text{Sorption capacity at RH=90\% (\%)} = \frac{W_{RH=90} - W_{RH=0}}{W_{RH=0}} \cdot 100$$

It can be observed that the obtained values are larger than those in the isotherms, meaning that the time necessary for reaching the equilibrium sorption is longer.

Table II.10. Water vapor sorption capacity and speed for a longer time (until sample weight remains constant at a relative humidity of 90%)

Sample	Weight at RH=0% (mg)	Weight at RH=90% (mg)	Sorption dynamic capacity RH=90% (%)	Time (s)	Sorption rate (10^{-3}%/s)
Reference 1	4.58	5.14	12.28	32	3.82
Sample 6	5.32	6.34	19.14	50	3.75
Sample 5	5.56	6.83	22.77	40	5.58

For the sample 5, the DVS analyses were made at two temperatures (25 °C and 35°C). In Figures 32 and 34 the influence of this parameter on the sorption/desorption isotherms and kinetics are presented.

Probably due to the hydrogen bonds formation favoring sorption, from Table II.10, the increasing of temperature leads to an increase in vapor sorption capacity of the sample.

Table II.11. Water vapor sorption capacity for sample 5 at both 25°C and 35°C respectively

Sample	Sorption capacity (%)
Sample 5_25	18.89
Sample 5_35	31.59

The differences between sorption-desorption speeds of those two temperatures for Sample 5 indexed sample are clearly highlighted by the presence of modified hysteresis and by the kinetics curves.

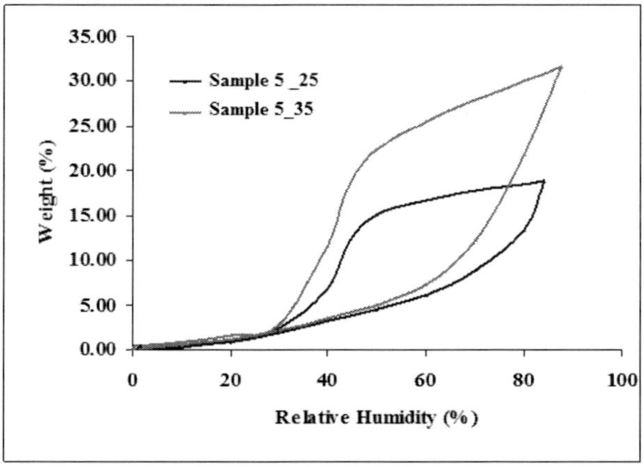

Figure 33. Comparative plots of rapid water vapors sorption/desorption isotherms for sample 5 at both 25°C and 35°C respectively.

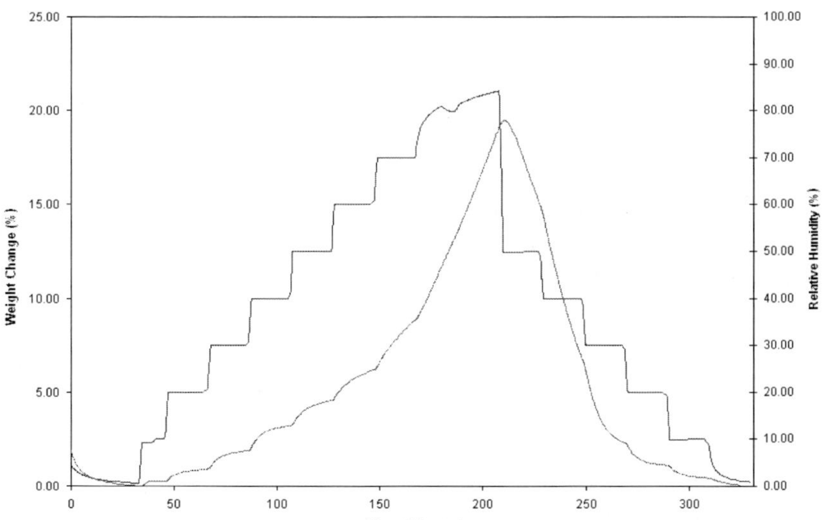

25°C

Figure 34. (Continued)

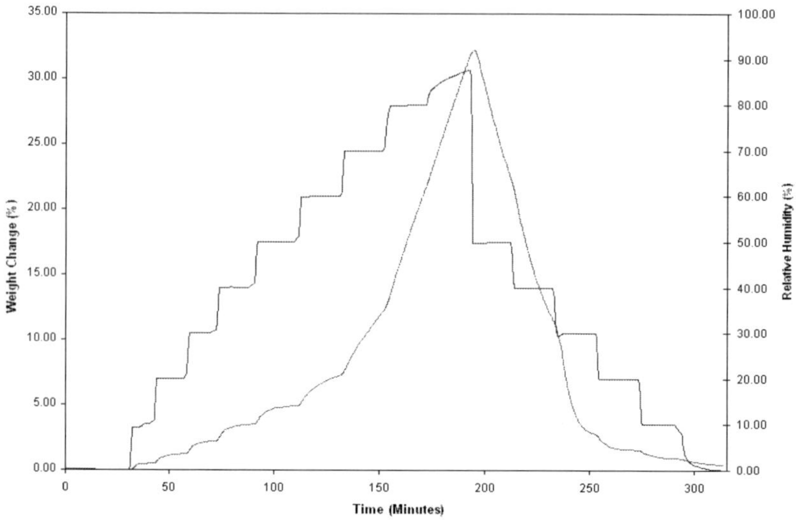

35°C

Figure 34. Kinetic curves for sorption/desorption processes of water vapors for sample 5 at both 25°C and 35°C respectively.

II.5. FT-IR Spectra Interpretation of Polymeric Fibrous Matrices Covered with Bulk/Micro and Nano ZnO

II.5.1. FT-IR Spectra/Chemical Modification of Hybrid Organic Inorganic Materials Made from In-Situ Synthesized Nano-ZnO Particles on Cellulosic Fibrous Matrices (Linen Supports), with the Assistance of CTAB and P123

Due to the stretching of the zinc and oxygen bond, in Figure 35, the *FT-IR* spectrum of hydrothermally synthesized, non-calcinated ZnO powder exhibited a high intensity broad band at about 430 cm^{-1}. A similar band was also observed in synthesized ZnO composites. As shown in the FT-IR spectrum of MCT-β-CD, the absorption bands between 1000 and 1200 cm^{-1} were characteristic of the – C –O– stretching on polysaccharide skeleton. In addition, two peaks appeared at 1420 and 1610 cm^{-1} corresponding to the symmetrical and asymmetrical stretching vibrations of the carboxylate groups

[51]. Moreover, the peak at 2920 cm^{-1} was ascribed to C–H stretching associated with the ring methane hydrogen atoms. A broad band centered at 3450 cm^{-1} was attributed to a wide distribution of hydrogen-bonded hydroxyl groups.

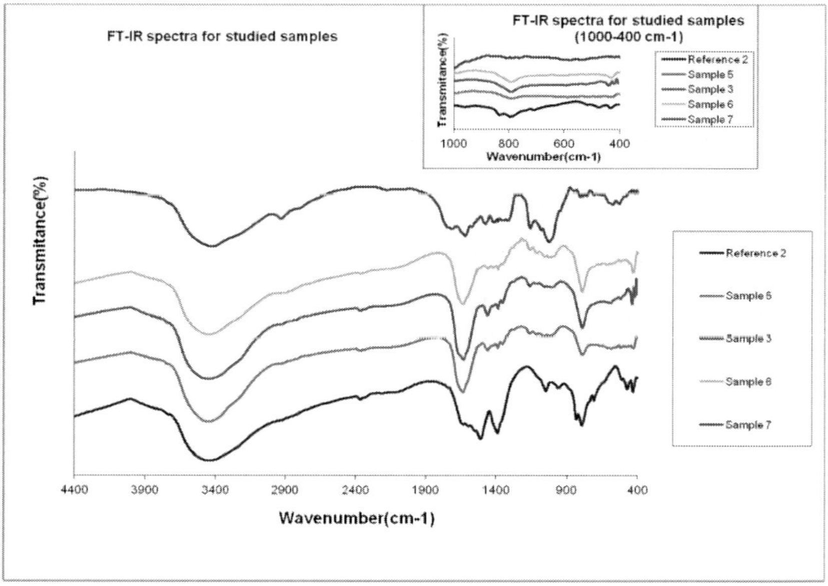

Figure 35. (a) FT-IR of spectra of: Sample 3; Sample 5; Sample 6; Sample 7; Reference [2].

The FT-IR spectra indicated that a strong interaction occurred in ZnO–MCT-β-CD nanoparticles, but no obvious formation of covalent bonds between MCT-β-CD and ZnO.

II.5.2. Chemical Modification /FT-IR Spectra of Hybrid Organic Inorganic Materials Made from *In-Situ* Synthesized Nano-ZnO Particles on Cellulosic Fibrous Matrices (Linen Supports), with the Assistance of MMT and MCT

Upon heating the change of IR spectra (Figure 36) of the nanocomposite shows the catalytic effect of ZnO. The absorption bands having a *shoulder* shape present in spectra at about 2914 and 2829 cm^{-1} wavenumbers corresponded to the methylene and methyl asymmetrical and symmetrical C-H stretching of aliphatic groups. The sharpness of these two bands is attributed to the low (weak) well-ordered, hydrocarbon chain. Corresponding to bond O-H (intermolecular hydrogen bond) vibration, the absorption band occurred in all the spectra at 3450 cm^{-1} wavenumber.

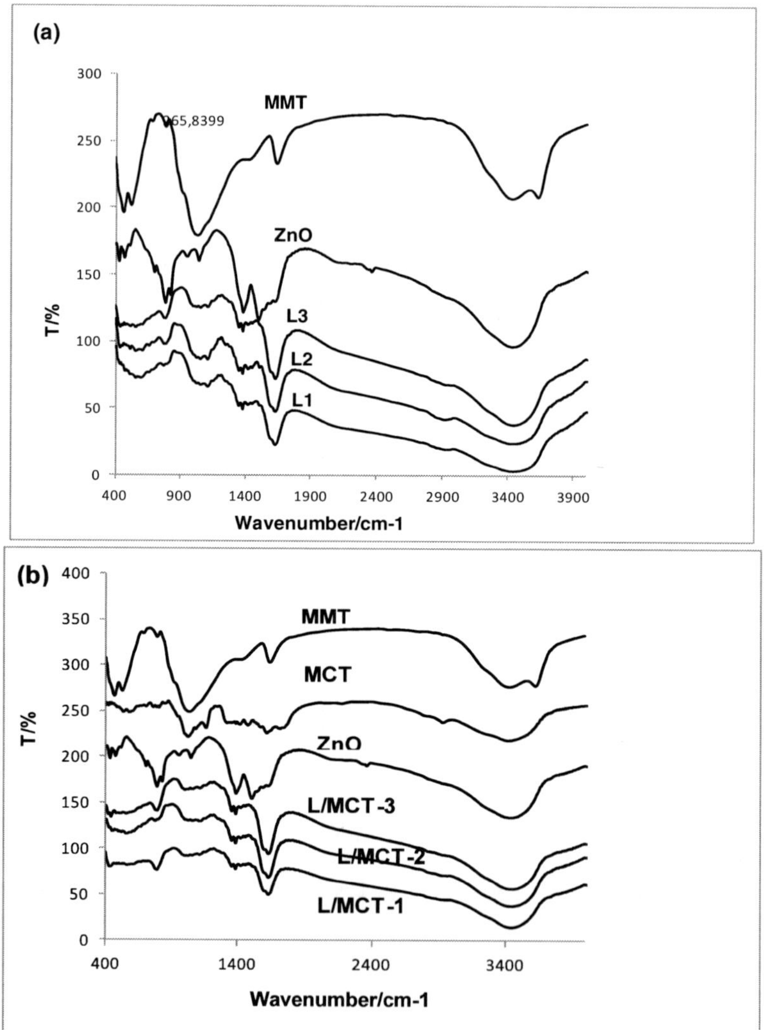

Figure 36. FT-IR spectra for: a) MMT, ZnO, L1, L2, L3 and b) L/MCT-1, L/MCT-2, L/MCT-3.

The 1335–1316 cm^{-1} doublet is assigned to the cellulose component. The absorption band from 1636 cm^{-1} can be assigned to C=C, aldehydic and carboxyl stretching vibrations.

As a subsequence of zinc ions in the reaction medium, the intermolecular association process is affected through the hydrogen bonds belonging to OH groups, since the zinc source (precursor) reacted with these groups, forming crystallization nuclei of zinc oxide. The deformation beyond the double bond plan from the lateral catena is highlighted in spectra, by bands occurrence at 790 cm^{-1}. The absorption bands present at 1115 cm^{-1} are related to the vibration-stretching mode C-O.

In case of montmorillonite, a band occurred at 3700 cm^{-1}, which could be assigned to non-hydrogen bonded O-H stretching vibration (Figure 36).

As a conclusion, the existence of ZnO nanoparticles prevents intermolecular hydrogen bonding, and therefore, the vibrational stretching modes of free O–H appear at a higher wave number.

II.5.3. Chemical Modification /FT-IR Spectra of Nano-Hybrids Made from Nano ZnO and Mono-Chloro-Triazinyl-β-cyclodextrin

By analyzing the significant changes in the shape and position of the absorbance bands of zinc oxide powder, monochlorotriazinyl-β-cyclodextrin, physical mixture and inclusion complexes, FT-IR technique allows the detection of complex formations in solid phase and points out the implication of the different functional groups of guest and host molecules in the inclusion process.

Figure 37 reveals some dominant features in the infrared spectrum of the ensemble of ZnO powders: Zn-O absorption band near 430 cm^{-1}. Theoretically, MCT exhibited significant FT-IR peak at wave number of 955, 1093, 1241, 1443, 2321 cm^{-1}, but in case of synthesized powder, the monochlorotriazinyl-β-cyclodextrin is shown by the 140 cm^{-1} vibration band corresponding to the -C=N- group from the triazinyc nucleus. By comparison with spectrum belonging to MCT-β-CD, the spectrum of ZnO embodied in MCT-β-CD, has the same bands, characteristic for triazinyc nucleus υ(C=N) meaning: at 1608, 1570 and 1471 cm^{-1} for MCT- β -CD. It is noteworthy that the band at 1570 cm^{-1} characteristic for triazinyc nucleus υ(C=N), decreased in intensity from the complex ZnO-MCT_4 - having the highest concentration of MCT, to ZnO-MCT_1 (the lowest concentration of MCT). This means that ZnO nanoparticles were indeed entrapped within the reactive matrix monochlorotriazinyl-β-cyclodextrin. The small shifts of the bands characteristic to triazinyc nucleus are due to the substituent modification (from -Cl in -O-R), following the reaction between monochlorotriazinyl-β-cyclodextrin and zinc nitrate. Here is a broad band with very low intensity at 3493 cm^{-1} corresponding to the vibration mode of water OH group indicating the presence of small amount of water adsorbed on the ZnO nanocrystal surface. The band at 1628 cm^{-1} is due to the OH bending of water. Moreover, intense peak at 1710 cm^{-1} (carbonyl group) is noticeable. The band at 500 cm^{-1} is attributed to the Zn-O stretching band which is consistent with that reported [169]. In case of the highest concentration of MCT ($ZnO-MCT_4$ compound), this band intensity is also the highest and becomes very low as intensity, in case of ZnO – MCT_1 complex. Therefore, it can be claimed that FT-IR spectra fills the results provided by the previous analysis, highlighting the incontestable role of MCT, in encaving the ZnO particles within polymer matrix.

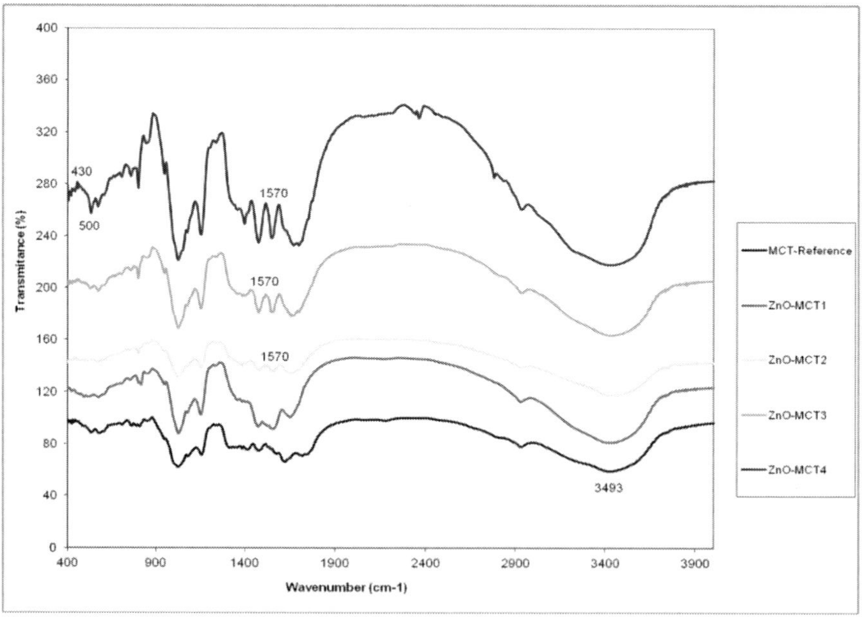

Figure 37. FT-IR Spectra of ZnO powders (ZnO-MCT1; ZnO-MCT2, ZnO-MCT3; ZnO-MCT4, compared with MCT (monochlorotriazinyl-β-cyclodextrin) spectrum.

II.6. Determination of the Comfort Features

II.6.1. Comfort Attributes for Hybrid Organic Inorganic Polymeric Composites Made of PES Fibrous Matrices Covered/Finished with Bulk/Micro Scaled/Powdered ZnO As Candidates to Antimicrobial Character [179]

The comfort features studied for samples prepared as presented in Table II.12 were: air permeability, water vapors passing resistance, porosity. Mean values of these parameters are shown in Table II.12.

Table II.12. Comfort features

Research route	δ [mm]	Pa [l/m²s]	Rv [mm*m²h/g]	Pz [%]	M [g/m²]
V0	0,234	559.039	0,0193	45,9	112,5
V1	0,234	558.835	0,0200	44,26	115
V2	0,234	398.378	0.0229	43,82	120,3
V3	0,234	397.441	0.0234	42,011	122,5
V4	0,234	396.098	0.0241	40,23	127

δ –thickness of studied samples.
P_a – air permeability.
R_v - the water vapors passing resistance.
P_z – porosity.
M – mass/surface unit (specific weight).

II.6.1.1. Air Permeability and Porosity Measurements

The values of the water vapors passing resistance, as well as the air passing permeability values are represented as histograms in Figures 38 and 39.

From the data presented in Table II.11 and Figure 38, the increase of zinc oxide concentration leads to a decrease of air passing permeability. This was noticeable at higher values of zinc oxide concentration: 5% and 7% respectively where Pa decreases from 558.835 l/m²s (value corresponding to 0-1% ZnO) to 397.441 l/ m²s, in case of 5% zinc oxide concentration, and to 396.098 l/ m²s for a concentration of 7% zinc oxide powder in the suspension.

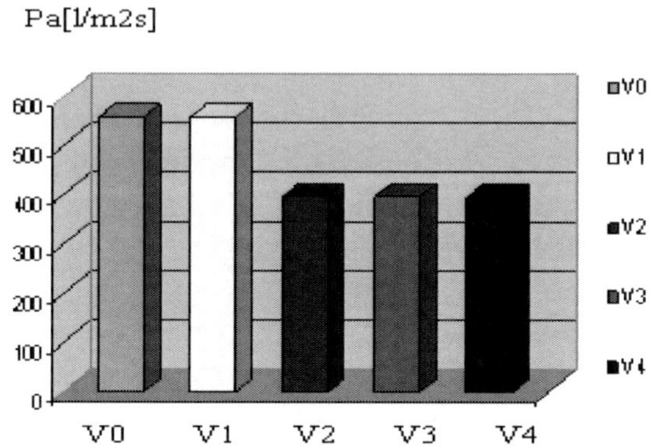

Legend: Pa [l/m²s] - air permeability; Rv [mm*m²h/g] - water vapors passing resistance.

Figure 38. Histogram of the air passing permeability.

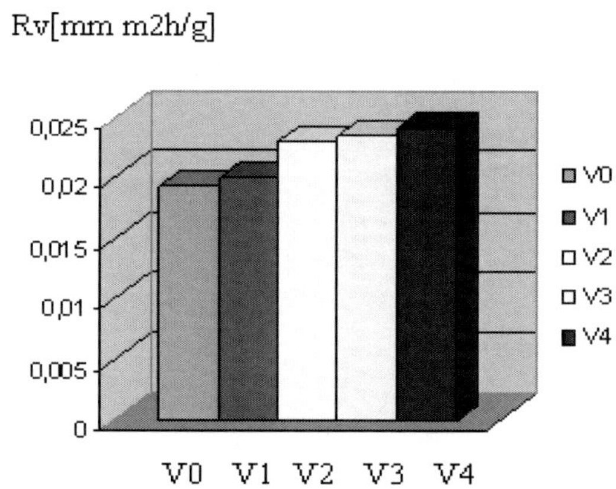

Legend: Pa [l/m²s] - air permeability; Rv [mm*m²h/g] - water vapors passing resistance.

Figure 39. Histogram of the water vapors passing resistance.

According to values from Table II.11 and Figure 39, it was noticeable that water vapor passing resistance increases when ZnO concentration was increased from 0.0200 mm×m²h/g at 1% ZnO concentration up to 0.0241 mm×m²h/g for 7% ZnO concentration.

The mass/surface unit M [g/m^2] increased with increase of the ZnO concentration, which is motivated by the fact that treatment with ZnO powder induced the covering of both fibril interstitials and pores fabric. These assertions are sustained also by the values obtained for porosity Pz [%], tending to decrease when ZnO concentration increased. Consequently, the V1 treatment route can be considered the most efficient due to the highest values of air permeability. In terms of water vapors passing endurance, the V1 treatment route leads to the best results, since water vapors passing endurance value was the lowest. This means that the water vapors transfer was performed in the easiest way possible [20].

II.6.1.2. Zeta Potential

The measurements of the Zeta Potential values before and after washing of the woven fabrics treated with ZnO are presented in Table II.13.

Table II.13. The Zeta Potential values before and after washing

Treatment type	pH	Zeta Potential		R^2	
		Before washing	After washing	Before washing	After washing
Binder + 1%ZnO	4	-48.5571	-18.9478	0.983	0.9
	7	-55.6872	-48.2668	0.993	0.81
	9	-64.5638	-40.8580	0.993	0.93
Binder + 3% ZnO	4	-45.9586	-21.0244	0.99	0.96
	7	-49.5885	-47.9903	0.99	0.89
	9	-43.6944	-52.4598	0.98	0.95
Binder + 5% ZnO	4	-33,.383	-27.8233	0.99	0.9
	7	-35.7732	-46.7724	0.77	0.92
	9	-38.1510	-43.1629	0.95	0.97
Binder + 7% ZnO	4	-18.2074	-20.6666	0.92	0.94
	7	-26.6908	-35.157	0.97	0.92
	9	-32.3889	-34.6020	0.96	0.98

As expected, the values of Zeta Potential after washing for specimens treated with ZnO at different concentrations, diminished compared to the initial values, as the Table II.12 shows.

Figures 40A and B presented charts showing Zeta Potential measurements before and after washing as a function of pH.

Analysis of the results presented in Table II.12 and Figure 40 lead to the following observations:

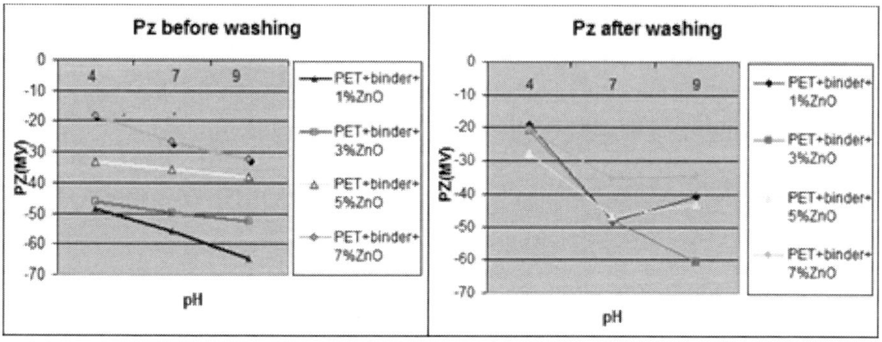

A B

Figure 40. The Zeta Potential (ZP) values as a function of pH and ZnO concentration before and after washing treatment; A-ZP before washing: B-ZP after washing.

a) Before washing: increasing the pH leads to more negative values of Zeta Potential for all ZnO concentrations. The increased ZnO concentration conducts to more negative values of zeta potential in all studied cases.
b) After washing: the overall pH dependency of Zeta potential remains the same but the electro-negativity increase became smaller.
c) The only ZnO concentration, which shows stability, proved to be 7%.

For all the studied specimens - it was noticeable - that Zeta Potential maintained its negative values. Moreover, these negative values for different concentrations of zinc oxide emulsion can be sustained by the dispersing agent's nature which is low anionic. The decrease of Zeta Potential can be explained by the fact that patterning of fabric surface was modified as the active agent concentration increases, in the detriment of the swelling capacity, as well as water sorption. The decrease of Zeta Potential at polyester fibers, in acid environment can be explained by the reducing of –OH and –COOH groups' dissociation, respectively [13, 111]. As a final conclusive remark, it can be claimed that washing treatment is recommended only in alkaline bath.

II.6.1.3. Wetting Properties. Contact Angle Measurements. Material Hydrophobicity

Upon examining Figure 41, comparing the three different experimental procedures of 100% PES fibrous, matrices with ZnO powder dispersed in methanol for 48 h had the strongest effect on CA values. The maximum CA value for the specific samples reaches 120°, meaning a hydrophobic surface, when the ZnO concentration in dispersion was 3%. Lowest values of CA are achieved for impregnation of 100% PES fibrous matrices with ZnO powder dispersed in water.

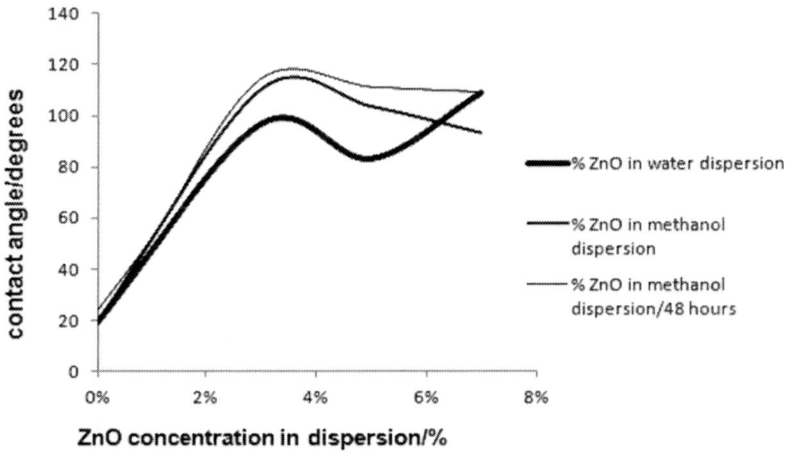

Figure 41. Contact angle variation for the various % ZnO in water and methanol respectively.

The largest CA value (about 100°) for these preparation conditions corresponds again to a concentration of 3% ZnO in dispersion. Impregnation of 100% PES fibrous matrices with ZnO powder dispersed in methanol reaches a maximum point (about 118°) at the same 3% ZnO concentration in dispersion. In any of the preparation conditions, these observations lead to the conclusion that 3% ZnO concentration corresponds to a maximum of the CA and best surface hydrophobicity. When using methanol as dispersion medium slightly higher values of CA with the best hydrophobicity for 48 hours-prepared samples were obtained. Higher ZnO concentrations lead to hydrophilic surfaces, as well as lower concentrations. These results are comparable with those presented by Popescu et al. for ZnO nanoparticles deposited by pulsed laser deposition method in optimized conditions [128].

II.6.2. Comfort Attributes for Hybrid Organic Inorganic Composites Made from PES Matrix and/Coated with/Covered with/Finished with Bulk/Micro Scaled/Powdered ZnO As Candidates to Anti-UV and High Hydrophobic Character

II.6.2.1. Wetting Properties. Contact Angle Measurements. Hydrophobic Responsivity

The material response to humidity and its wettability control the material performance in presence of water, vapors or humid environment. This performance relates both to the degree of comfort of the textile and to its wear behavior, since a lower wettability/higher hydrophobicity surface is producing a higher self-cleaning efficiency. In this respect, related to decreased permeability to air and increased resistance to vapors, after ZnO impregnation, the air permeability Pa and the resistance to vapor transfer Rv, presented in Table II.12, are modified. Generally, Pa and Rv values follow the trend of evolution of the ZnO concentration of the finishing agent. The relative variation of these parameters, compared to the uncoated sample, ΔPa/Pa and ΔRv/Rv, is also presented in Table II.14. This information shows that the air permeability, which already diminishes for 1% ZnO sample, is strongly reduced for the 3% ZnO sample, and then levels out for higher amount of zinc oxide in the finishing agent. It is noteworthy to mention that constant Pa values are obtained for 3–7% ZnO impregnated fibrous matrices. Rv responses somehow similarly, meaning that the initial increase observed for the 1% ZnO sample is followed by a strong increase for the 3% ZnO textile. For ZnO amounts higher than 5%, the evolution seems to stabilize. After UV irradiation, measurement of Pa and Rv does not show measurable variation, stressing the stability of the ZnO-modified material. Thus, coated polymeric matrices offer enhanced hydrophobicity and better protection against contacting gases and humidity. In order to improve comfort of polyester fibrous matrices, without overloading the fabric a 3–5% ZnO level of impregnation of the matrices yields stable properties. The matrices' wettability of the fabric, evaluated by contact angle measurement, is presented in Table II.15. As an indicator of hydrophobic character, the measured water contact angle on the uncoated material is 91°. This value is higher than that generally reported, of about 80° on smooth PET films [177, 126], and could relate to the intrinsic rough surface of the woven material. Visibly, the impregnation with ZnO conducts to higher hydrophobicity, since significantly increases on coated samples. This increase is too large to be related to the rough surface of the fabric, which is comparable for uncoated and coated samples. Besides, the measurement was repeated on various sites of each sample, the contact angle values showing limited dispersion (±2°) for all materials, with a symmetrical profile of the drops, which points to satisfactory homogeneity of the coating. In this respect, Figure 42 shows the image of a water drop on the 3% ZnO coated polyester, where the measured contact angle is 135° [21].

Table II.14. Air permeability and resistance to vapor transfer R_v , for polyester fibrous matrices coated with ZnO

Sample	P_a (l/m²s)	R_v (mm·m²h/g)	$\Delta P_a / P_a$ (%)	$\Delta R_v / R_v$ (%)
polyester	519.4	0.0188	—	—
polyester + 1% ZnO	558.8	0.0200	-7.5	6.4
polyester + 3% ZnO	398.4	0.0229	-23.2	21.8
polyester + 5% ZnO	397.4	0.0238	-23.5	26.6
polyester + 7% ZnO	396.1	0.0241	-23.7	28.2

Figure 42. Photo of a 1 μl water drop on the 3% ZnO coated polyester fibrous matrix (135° contact angle).

Our samples present a high degree of hydrophobicity, demonstrated by the water drops deposited onto the surface vanishing after long intervals, by evaporation. Furthermore, there is no deformation of the drops due to preferential direction along fiber or yarn orientation. The drop contour, seen from the top, is circular, and the drop profile, seen from the front, is symmetrical (see Figure 42). We, therefore, consider that this measurement yields correct and reproducible information to compare the coated and uncoated samples.

Table II.15. Contact angle and adhesion work of water on polyester fibrous matrices coated with ZnO, before and after UV irradiation

Polymeric fibrous matrix	before UV irradiation		after UV irradiation	
	θ (°)	W_a (mJ/m^2)	θ (°)	W_a (mJ/m^2)
polyester	95°	66	86°	78
polyester + 1% ZnO	121°	35	117°	40
polyester + 3% ZnO	134°	22	130°	26
polyester + 5% ZnO	128°	28	126°	30
polyester + 7% ZnO	119°	37	113°	44

The diminution of the adhesion work Wa, with an average 50% leads to the increase in contact angle. This variation is explained by the presence of the ZnO layer onto polymeric fibrous matrices, where, the oxide layer is more hydrophobic than the polymer, which is in agreement with other measurements reported in the literature [16].

Noteworthy, the lowest wettability is observed for 3% ZnO impregnated fabric, whereas increased ZnO concentration leads to higher wettability, i.e. reversal of the surface properties. A very relevant aspect is that between 3% and 7% ZnO samples a 68% difference in adhesion work occurs. This response relates to better dispersion of ZnO particles in the 3% sample, which is more diluted, allowing more uniform coating of textile fibers, and more hydrophobic character due to oxide layer. In case of specimens impregnated at higher oxide concentration, the agglomeration of ZnO particles conducts to non-uniformity, and therefore the matrices are irregularly coated with layers of different thickness.

As expected, the UV irradiation of the samples conducts [16] to diminution of the contact angle (Table II.15). Given the surface modification of polymers under exposure to energy sources, the diminution is visible for the polyester fibrous matrix. Thus confirming the UV protective function of the oxide layer, the modification is much less important for the ZnO coated fibrous matrices. After UV exposure, all ZnO coated samples retain a pronounced hydrophobicity. The modification has the highest degree for the 7% ZnO sample. Once again, it is confirmed that a better dispersion of the microparticles and the good uniformity of the surface layer, ensured for 3–5% ZnO solutions, relate to enhanced UV-protection.

In order to test the stability of the coating to washing, contact angle was also used, stating that the material resists without significant modification to five washing cycles, which is reasonable, since this material could find applications for outdoor activities. The material resists very well to the contact with water, its hydrophobic properties do not reverse.

Thus, this polymeric fibrous hybrid, obtained by an alternative approach to the usual nanoparticle impregnation or coating methods, would combine time stability, water repellent and UV blocking properties.

II.6.2.2. Electric Attributes

The performance of comfort of the polymeric fibrous matrices is in direct relation to the *electrostatic charging*; consequently, the *electric conductivity* is quantifying the material ability to neutralize the electric charge present in the environment. Since UV irradiation may induce supplementary charging of dielectric materials, which is affecting again the degree of comfort, the assessment of the dielectric properties of the material is essential.

Being dependent on the mechanical and chemical properties of fibrous polymers, crystalline-to-amorphous ratio, moisture content and dyes, impurities, static charges, etc. the polymeric fibrous matrices are usually dielectric materials, characterized by high electrical resistance. The occurrence of ZnO coating on polymeric matrices is expected to modify the dielectric characteristics of the polyester material.

The evolution of the electrical conductivity σ of polyester fibrous matrices, upon addition of ZnO, compared to the reference sample, in the 1–106 Hz range is presented in Figure 43. Although the difference is small, from 10–12 to 4 × 10–12 S/m (at 10 Hz), the contribution of the ZnO conductivity is visible in the low frequency range (inset Figure 43), which corresponds to the orientation of permanent dipoles, and is in agreement to the composition of the coated material. Importantly, the smooth evolution of σ denotes good uniformity of the samples [21].

Showing the dominant dielectric capacity of the polymeric substrate within the organic–inorganic material, *the complex impedance Z* characteristics, represented on Cole–Cole diagrams, are similar, within error bars, for the textile samples with different % of ZnO. An apparent single component in impedance illustrates a rather homogeneous defect distribution in textile samples, without any sign of a "brick-wall" electric microstructure. As a conclusive remark, the ZnO addition does not modify the internal resistivity of the samples and, importantly, does not induce any electric inhomogeneities. In terms of electric properties, the ZnO coating, in the 1–7% ZnO range, shows satisfactory uniformity.

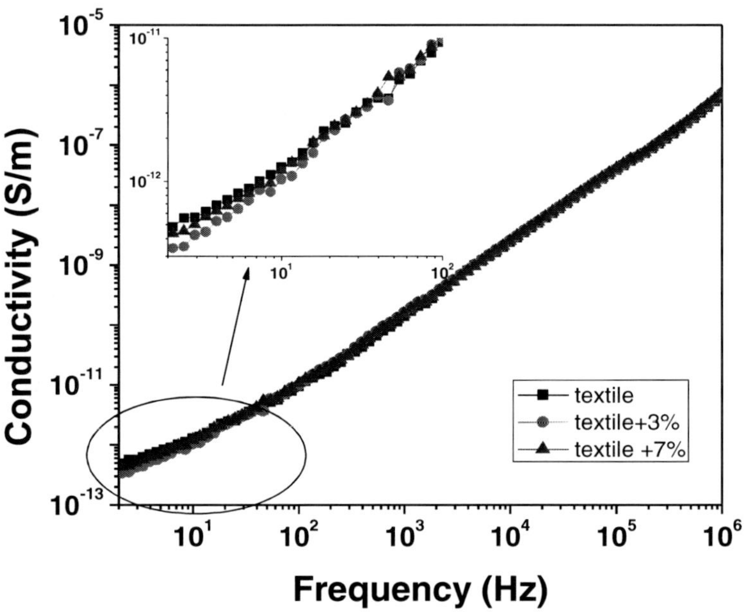

Figure 43. Conductivity vs. frequency for polyester fibrous matrices coated with ZnO.

II.6.2.3. Material Response to UV Radiation

Due to microparticles and taking into consideration that ZnO is recognized as UV absorber, yet having confined local action, the overall response of the material has to be monitored after systematic UV irradiation, under controlled conditions.

The absolute reflectance of the material was firstly measured, in this respect. The absolute reflectance, presented in Figure 44, has low values for all samples, due to the absorbent character of the polyester fabric. Nonetheless, the relation to the ZnO coating, which is enhancing the fabric absorbance, is visible.

By comparison the reference material (not impregnated polyester), indicating a higher UV absorbance, in the UV 250–350 nm range, the coated material has lower by 50% reflectance. In this range, there is no measurable difference between samples with different ZnO concentrations, all showing comparable values of the reflectance.

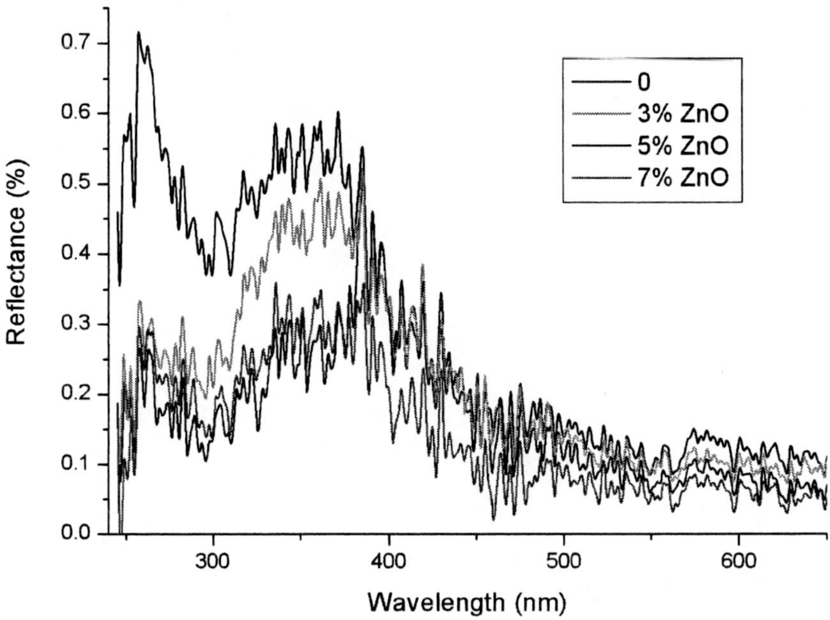

Figure 44. Absolute reflectance of the polyester fibrous matrices coated with ZnO, in the UV-Vis range.

In the 300–400 nm range there is a difference between the coated fibrous matrices, where the 3% ZnO sample has significantly lower reflectance. For materials with 5% and 7% ZnO, the reflectance levels out. Although the general trend of variation of the UV reflectance respects the order of the concentrations, i.e. 0, 3%, 5%, 7%, results suggest that an optimum concentration related to UV absorbance corresponds to 3–5% ZnO.

To compare the dielectric character of the samples after UV irradiation, the Cole–Cole diagrams are employed. The impedance diagram for the textile samples, where the convention of representation places the low frequencies in the top-right corner and the high frequencies in the bottom-left corner is showed in Figure 45. As a correlation between the dielectric characteristics of the fibrous matrices and the UV treatment should be stated: the dielectric attributes are perturbed by UV radiation.

Depending on the ZnO impregnation of the fibrous matrices, there is an obvious different behavior of the samples. In terms of polar group orientation, for all specimens, for the low frequency range, differences in the dielectric character occur. In the high frequency range, concerning only the displacement polarization, all curves converge.

The stability function with time of the samples, against UV radiation, is tested by measurement at different intervals after the UV exposure. Consequently, the uncoated polyester fibrous matrix is the most strongly perturbed by UV, showing increased dielectric character, which remains practically unchanged even one week after irradiation.

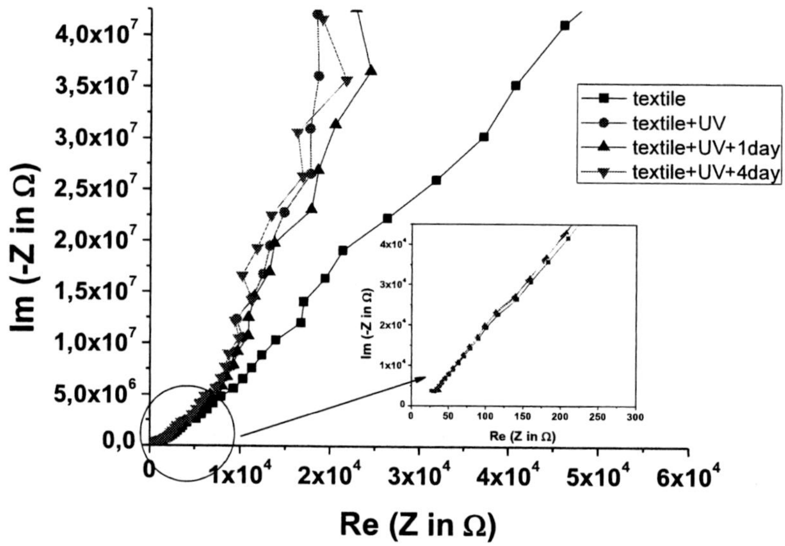

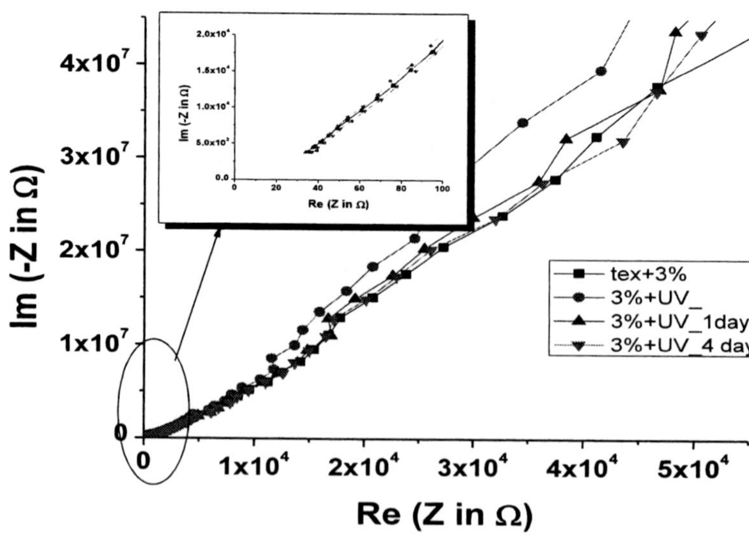

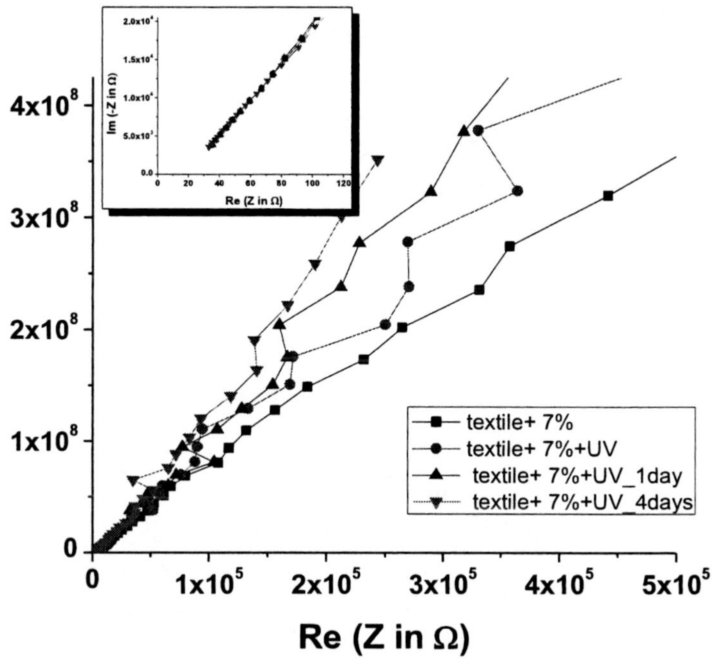

Figure 45. Cole-Cole diagrams for polyester fibrous matrices: a) uncoated, b) 3% ZnO, c) 7% ZnO.

The UV exposure of the polymer conducts to increased surface charging, which reduces the degree of comfort of the textiles. The 3% ZnO coated polyester shows much smaller perturbation of its dielectric characteristic by UV exposure, which, importantly, already reverses to the untreated material one day after irradiation.

The uniform ZnO coating of the polymer is playing a fundamental role in the process of a better absorption of UV and dissipation of surface charge, enhancing the comfort and quality of the garment. As the initial UV-induced variation of the dielectric character further amplifies during the days following irradiation, the 7% ZnO coated sample shows unstable responsivity. For these samples, the process cannot be controlled and could be due to the nonuniform surface layer, where agglomerations of ZnO alternate with polyester, the two components bearing different dielectric functionality. As a conclusion, the most stable material seems to be the 3% ZnO impregnated fibrous matrix, as the dielectric character and the surface charging, modified by UV exposure, are reverse to those of the unexposed material and remain stable for several days after irradiation.

II.6.2.4. Mechanical Properties

The evaluation of the mechanical properties shows that the fibrous matrices are not affected by ZnO impregnation. Thus, the textile strength is about 670 MPa, having similar values, within error bars (±5%), for coated and uncoated samples. Moreover, the mechanical stability indicates no modification after UV irradiation.

II.6.3. Thermal Properties

II.6.3.1. Thermal Degradation of Hybrid Organic Inorganic Materials Made from In-Situ Synthesized Nano-ZnO Particles on Cellulosic Fibrous Matrices (Linen Supports), with the Assistance of MMT and MCT

Since the results provided by the XRD analysis are not very relevant, they can be completed, correlated with the DSC studies, since the last type of investigation is able to evaluate the crystallization/melting processes [183]. The DSC curves of hybrid materials under nitrogen are shown in figures above. Generally, in a typical DSC curve of cellulosic fibrous matrices, there is an endothermic peak in the range 370°-395°C, which has been shown to be primarily due to the production of laevoglucosan [208].

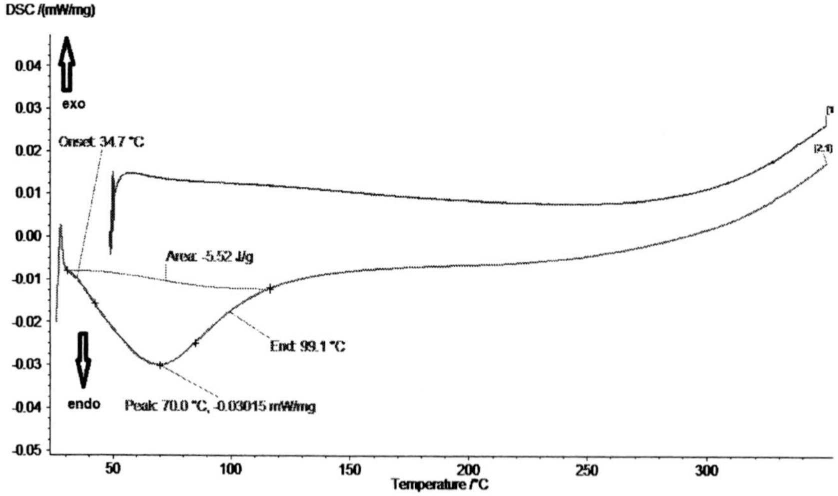

a

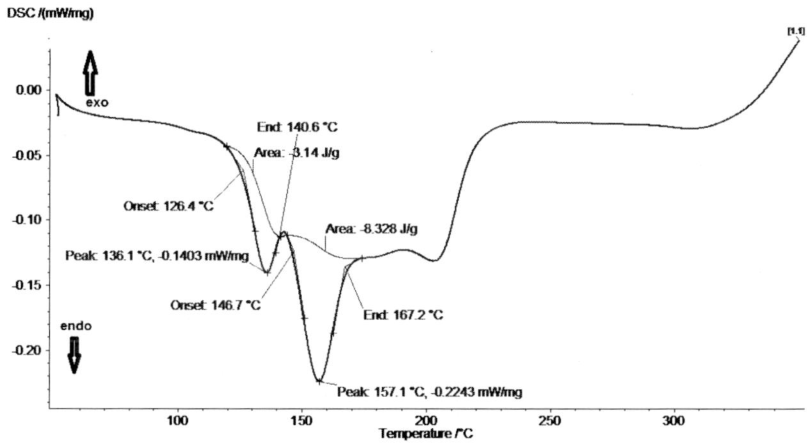

b

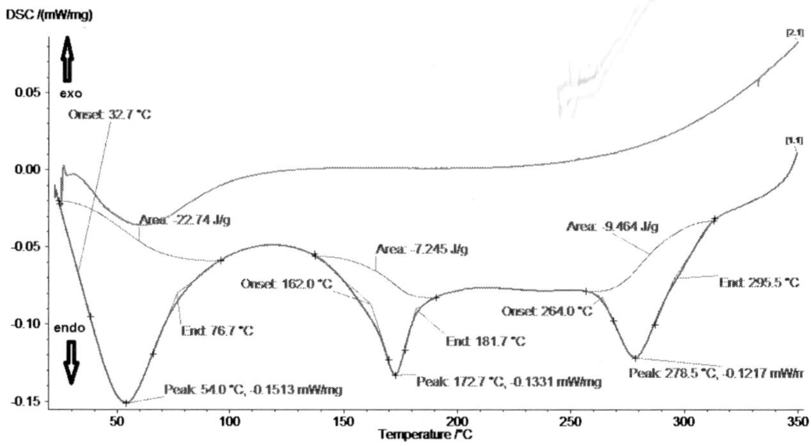

c

Figure 46. Typical DSC curve under nitrogen for: **a** L1; **b** L2; **c** L3.

Particularly, for linen fibers, this peak is sometimes partly or completely marked by an exothermal effect around 340°C, attributed to a base-catalyzed-dehydration reaction that takes place in the presence of alkaline ions, such as those of sodium [138].

It is well-known from the literature that lignocellulosic fibers degrade in several steps. A progressive mass loss was observed from 200 to 250°C associated with water release. The hemicellulose degrades at about 240°–

310°C, whereas the cellulose degrades between 310°–360°C and the lignin has been shown to degrade in wide temperature interval (200°–550°C) [113].

It is remarkable that the nanocomposite treated with ZnO nanoparticles with the assistance of MMT started to decompose at higher temperature than sample treated in the same conditions but without the presence of zinc oxide.

It is not possible to separate the different degradation processes of the fiber components because the reactions are very complex and overlap in the range of 220°–360°C.

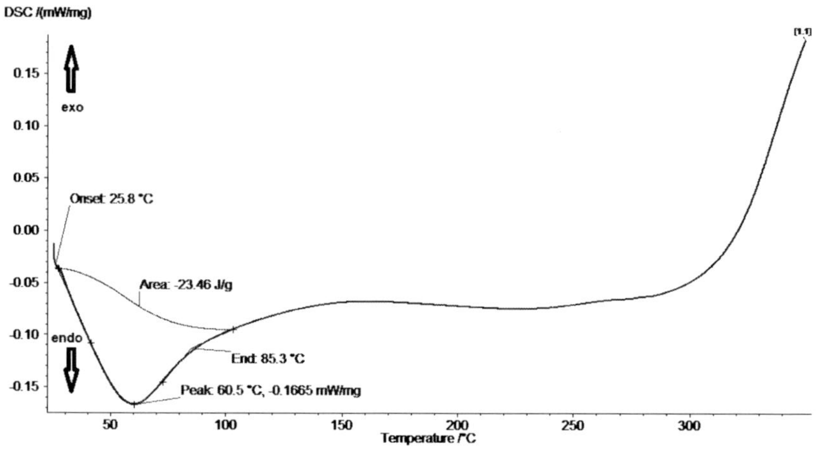

a

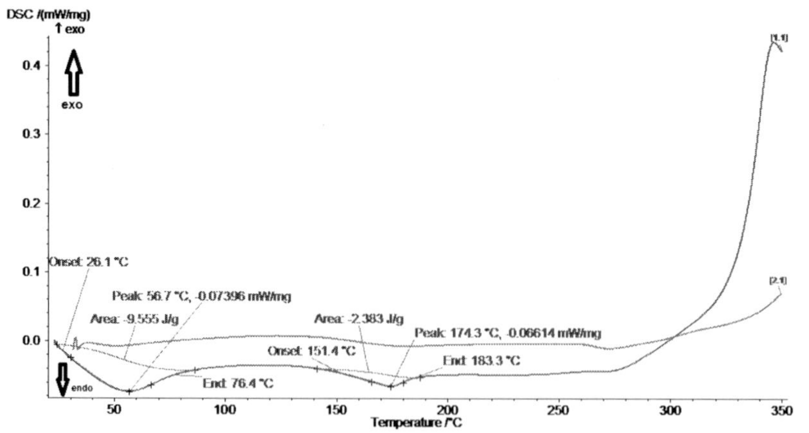

b

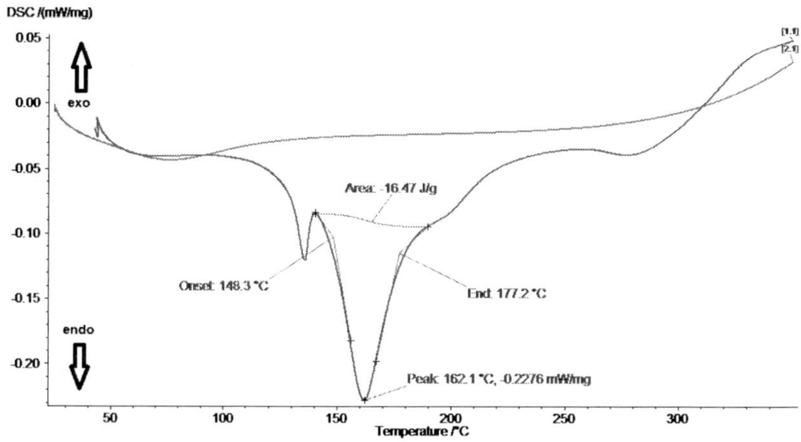

c

Figure 47. Typical DSC curve under nitrogen for: **a** - L/MCT-1; **b** - L/MCT-2; **c** - L/MCT-3.

However, the existence of the montmorillonitic clay on the surface of the probes delayed the thermal degradation of the cellulosic fibrous matrices (linen samples), even those non-treated with zinc oxide particles.

Cellulose is thermally decomposed through two types of reactions. At lower temperatures, there is gradual degradation which includes dehydration, depolymerization, oxidation, evolution of carbon monoxide and carbon dioxide, and formation of carbonyl and carboxyl groups, ultimately resulting in carbonaceous residue forms.

In Figure 46 (a) and (b) the DSC curves show an endothermic band around 260°C, indicating a weight loss. In Figure 46 (a) as the temperature rises to 310°C, the surface acidity of zinc oxide nanoparticles keeps accelerating the decomposition of the fibrous matrix. Thus, a very much lower amount of carbonyl groups is found in the linen - ZnO nanocomposite specimens, according to the FT-IR spectra. Having a higher thermal conductivity as well as a greater heat capacity value, MMT absorbs the heat transmitted from the surroundings and retards the direct thermal impact to the polymer backbone [89, 142]. As a result, zinc oxide stabilizes the polymer molecules of the underneath substrates and delays the occurrence of major cracking up to 400°C (Figure 47). The response of the reference fibrous cellulosic (non-functionalized) subjected to the thermal treatment in N_2 clearly shows the masking effect of an exothermal reaction on the endothermic cellulose decomposition.

Surprisingly, even within the second cycle of thermal treatment, the sample exhibits a similar exothermal peak at 363°C. The sample shows an exothermal peak at 260°C with a decreased enthalpy after the thermal treatment; the exothermal effect is attributable to β-cellulose decomposition as observed in a curve of a cotton sample.

II.6.3.2. Thermal Degradation of Hybrid Organic Inorganic Polymeric Composites Made of PES Fibrous Matrices Covered/Finished with Bulk/Micro Scaled/Powdered ZnO As Candidates to Antimicrobial Character

In order to determine the thermal behavior of hybrids their thermal decomposition was examined by using TGA under nitrogen environment. Figure 48 (a and b) shows the TGA curves of *hybrid organic inorganic polymeric composites* prepared from the various ZnO concentration dispersions. For the specimens resulted from water dispersions the TGA curves showed that the untreated PES fibrous matrix decomposes at a maximum rate at around 370°C (nitrogen environment) as expected based on the fact that PES fibers displays good heat resistance. ZnO presence onto matrices leads to a dramatic decrease of decomposition temperature, the sample prepared from 5% ZnO in dispersion beginning decomposition at 150°C. This may advance the observation that the coating finishing with ZnO does not improve the thermal stability of PES fibrous matrices but the opposite. The polymeric fibrous matrices treated with the highest ZnO concentrations (samples 5, 4 and 3) showed a two-stage decomposition pattern which became more obvious at higher concentrations of ZnO.

The material decomposition occurs in two major steps separated by a development of a residue around 470°C, which progressively and totally degrades up to 560°C, in accordance with similar studies [31]. Particularly, in our study, the first initial decomposition between 120°C and 400°C could be explained by the thermal decomposition due to random scission of the ester linkage and pyrolysis of the simple esters. Following the second stage of decomposition between 400 and 470°C, newly formed end groups decompose by secondary reactions [53].

The reference PES sample is thermally stable up to 380°C, temperature at which the neat thermoplastic loses 5% of its weight. For the second stage decomposition between 400 and 470°C, another possible explanation, the reduction of Zn in the composite may be accentuated by increasing ZnO concentration.

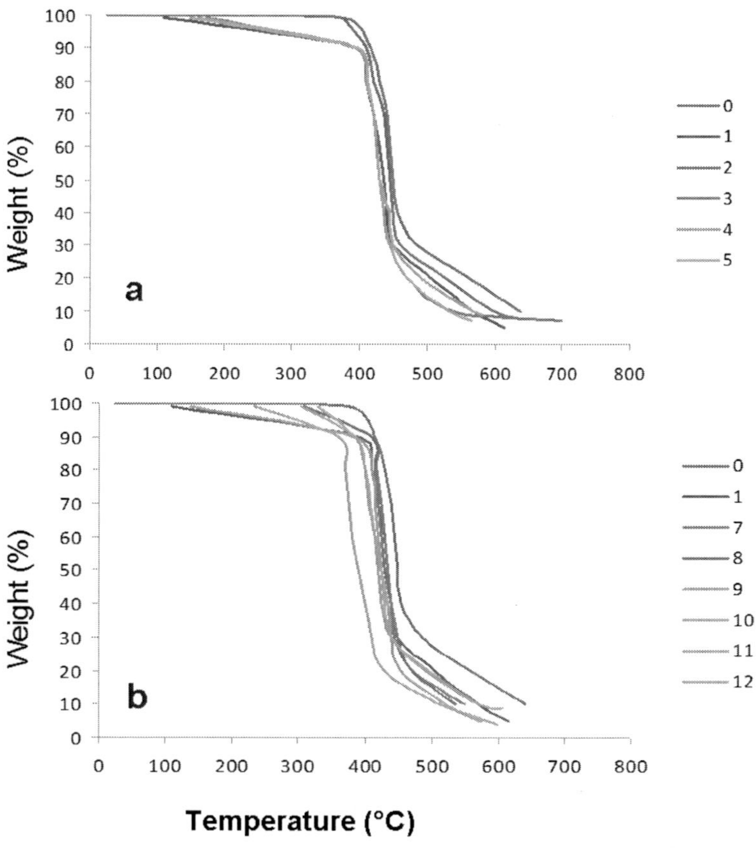

Figure 48. TGA curves of hybrid organic inorganic polymeric composites prepared from the various ZnO concentration dispersions: a) in water, b) in methanol.

The decomposition temperature (Td) values of the 100% polyester fibrous matrix treated with 3%, 5% and 7% concentrations of ZnO are 230°C and 415°C, respectively, meaning much lower than the untreated samples. Subsequently, the treatment of PES fibrous matrix with ZnO induced a decrease of thermal stability of polyester support and the stability diminished with the increase of ZnO powder concentration. Regarding the coated fibrous matrices, hybrid organic inorganic polymeric composites reveal a thermal resistance at high temperature starting to degrade at 370 up to 460° C where the material decomposition is slowed by a stabilization [183].

This may be associated with ZnO presence as oxygen source for the PES fibrous matrices' thermal degradation. It is noticeable that, according to graph plotting of the thermogravimetric curves (Figure 48 a and b), the control sample is the most stable with respect to thermal decomposition. The thermal stability of hybrid organic inorganic polymeric composites seems to be almost the same in both cases of ZnO dispersed in water and methanol (short and 48 hours). The "highest" Td among these (being thermally stable up to 300°C) appears to be the samples denoted **8**_PES + 5%ZnO + methanol, **10**-PES + 1%ZnO + methanol/48 hours and **12**_PES + 5%ZnO+methanol/48 h.

In: Novel Approach of Added-Value Zinc Oxide ... ISBN: 978-1-63483-432-2
Editors: N. Vrinceanu and P. Postolache © 2015 Nova Science Publishers, Inc.

Chapter 3

CONCLUSION

Narcisa Vrinceanu,* PhD
"Lucian Blaga" University of Sibiu, Romania

1. HYBRID ORGANIC INORGANIC MATERIALS MADE FROM IN-SITU -SINTHESIZED NANO-ZNOPARTICLES ON CELLULOSIC FIBROUS MATRICES (LINEN SUPPORTS), WITH THE ASSISTANCE OF CTAB AND P123

Part of the research was conducted on a series of MCT-β-CD grafted polymeric fibrous matrices (linen fibrous support - yarn) in whose matrices zinc oxide nanoparticles were introduced with the assistance of two different surfactants. The coating particles fell off easily as the ZnO powder hydrothermally synthesized without any surfactant assistance after washing, which might have been caused by the weak attaching force (coordinated bond between ZnO and linen) induced by the deteriorated crystallinity.

Wetting characteristics are influenced by the type of surfactant used during the hydrothermal synthesis, with direct impact on the relationship between the morphological, structural and chemical attributes and water vapor sorption-desorption behavior. The hydrophilicity of these modified fibrous matrices increased. Based on the sorption/desorption isotherms registered by DSV, BET surface area, correlated with results provided by XRD analysis,

* Corresponding author address. E-mail: vrinceanu.narcisai@ulbsibiu.ro.

these fibrous hybrids were evaluated and assimilated, with mesoporous materials, in accordance with IUPAC regulations,. The humidity loss and drying speed of water from these studied samples depend on the type of surfactant.

2. Hybrid Organic Inorganic Materials Made from In-Situ Synthesized Nano-ZnO Particles on Cellulosic Fibrous Matrices (Linen Supports), with the Assistance of MMT and MCT

On the other side, the study involving in thermal responsivity emphasized some essential aspects, like:

- A novel preparing technique in order to obtain nanoscale ZnO coated fibrous hybrids, by using this time the synergic effect of another two main components, MMT and MCT;
- The thermal stability and degradation mechanism of ZnO nanocoated cellulosic fibrous matrices;
- The cumulative barrier attributes conferred by the two new components that interfered in the preparation technique: MMT and MCT.

From the SEM photos and by comparing the morphology of the two types of samples before and after thermal treatment, some randomly distributed conglomerations are noticeable, in case of *non-functionalized sample*, while on the *functionalized cellulosic polymeric sample*, the particles have bigger dimensions uniformly distributed.

EDX investigation highlighted a decreasing of zinc oxide content with the increasing of the treatment temperature, for both studied samples.

However, the XRD analysis revealed the fact that the entrapment was not good-enough:

- In case of functionalized samples set, CI increases with the decreasing of the crystallites nanoparticles;
- In case of non-functionalized cellulosic fibrous matrices, there is a proportionally increasing of CI with the augmentation of nanoparticles size.

- The Differential Scanning Calorimetry (DSC) also revealed higher enthalpy value compared to the reference one. The synthesized nano-composites revealed higher thermal stability and higher enthalpy value compared to the reference one. In the meantime, MMT having a higher thermal conductivity as well as a greater heat capacity value, absorbed the heat transmitted from the surroundings and retarded the direct thermal impact on the polymer backbone.
- Chemical composition and crystallinity of the studied samples influenced the thermal degradation behavior of the supports studied. Thus, a very much lower amount of carbonyl groups is found in the linen - ZnO nanocomposite specimens, according to the FT-IR spectra. FT-IR spectra revealed the existence of zinc oxide and MMT.
- The perspective research will be oriented to a more improved technique of fixing the ZnO nanoparticles in order to add multifunctionality of the studied fibrous nanocomposites.

3. Nano-Hybrids Made from Nano ZnO and Mono-Chloro-Triazinyl-B-Cyclodextrin

A novel and simple methodology of growing the ZnO–monochlorotriazinyl-β-cyclodextrin nanocomposites via a facile route was developed. The synthesis showed a lamellar ZnO nanostructure, due to the assistance of monochlorotriazinyl-β-cyclodextrin, using nitrate as crystal growth modifier. The study revealed a surprisingly structural similitude between ZnO nanoparticles dispersions, using the polymer matrix, and natural biocomposites.

A co-assisted system studied the microstructures: X-ray diffraction method, scanning electron microscopy, and FT-IR spectroscopy techniques. In the nucleation process an essential role was played by monochlorotriazinyl-β-cyclodextrin–ZnO interaction to the formation of complex structure, forming the high-quality lamellar nanostructures. The morphology, dimension, and size distribution of the product ZnO are strongly affected by the presence of monochlorotriazinyl-β-cyclodextrin.

The SEM figures suggested that a two-length-scale hierarchical structure is formed on the surface. The nanoparticles are well dispersed in the MCT nanocavities, although some aggregated nanoparticles are still visible. The particles size plays a primary role in determining their adhesion on a hypothetic/potential fibrous matrix. This kind of structures is in accordance with water repellent properties exhibited by the lotus leaf. Textural information used in this research can be used efficiently for precise structural determination of new synthesized compounds made of ZnO structures onto fibrous supports, in order to achieve enhanced attributes of hydrophobicity and selfcleaning.

The study can ensure the fundamentals for a new one in which for ZnO nanoparticles entrapped in MCT (monochlorotriazinyl-β-cyclodextrin), guest molecules can confer high durability, both for ZnO nanoparticles dispersions and treated fibrous matrices, as compared with conventional materials. They possess a large surface area and high surface energy that ensure better affinity for fibrous matrices and lead to an increase in durability. Wash fastness/stability of the nanoparticles is a particular requirement for textiles and it is strongly correlated with nanoparticles adhesion to the fibers. Wash fastness can be further improved with the formation of covalent bonding between nanoparticles and the fibrous matrices surface.

In order to achieve selective growth of a desired morphology, understanding the nanostructure morphology can be considered assistance in monitoring fabrication processes. In return, unique quantum confinement effects can be maximized from the desired crystallographic orientations, which in the future will lead to prototype novel nanodevices of important commercial value.

The results of the present work could be leveraged in the area of consumer textile due to the adjusting to a desired textile nanocomposite with an architecture allowing blocking/fixing/adhesion of different guest compounds, with different properties, such as water repellence/washing resistance, antimicrobial activity and photoprotection.

4. Hybrid Organic Inorganic Polymeric Composites Made of PES Fibrous Matrices Covered/Finished with Bulk/Micro Scaled/Powdered ZnO as Candidates to Antimicrobial Character

In terms of comfort associated parameters as well as Zeta Potential as a measure of treatment stability, various concentrations for the treatment of polyester fibrous matrices were studied.

This study leads to the following conclusions:

- The increase of zinc oxide concentration leads to a decrease of air passing permeability from 558.835 l/m²s (value corresponding to 0-1% ZnO) to 397.441 l/ m²s, in case of 5% zinc oxide concentration, and to 396.098 l/ m²s for a concentration of 7% zinc oxide powder in the suspension.
- Water vapor passing endurance increases when ZnO concentration was increased from 0.0200 mm×m²h/g at 1% ZnO concentration up to 0.0241 mm×m²h/g for 7% ZnO concentration.;
- The mass/surface unit M [g/m2] increased with the increase of the ZnO concentration. This fact can be motivated by the fact that treatment with ZnO powder induced the covering of both fibril interstitials and pores fabric.
- These assertions are sustained also by the values obtained for porosity Pz [%], tending to decrease when ZnO concentration increased. Comfort associated parameters values decrease with the increase of ZnO concentration. According to these results, the best comfort parameters values correspond to 1% ZnO concentration treatment.

Zeta Potential values, belonging to specimens treated with zinc oxide suspensions with different concentrations, before and after washing, decreased with the increase of zinc oxide concentration. However, stability was noticeable for samples corresponding to the highest ZnO concentration (7%). After washing, Zeta Potential values of ZnO suspension treated samples, in different concentrations, decrease with the increase of ZnO concentration in the treatment dispersions. The diminishing of Zeta Potential values in case of polyester fibers, when the washing medium becomes more acid, can be explained by the reduction of the dissociation of –OH and–COOH groups. The best stability was shown by sample resulted from 7% ZnO treatment. This suggests that a compromise between comfort and ZnO treatment stability for improved antimicrobial properties should be considered. Further studies are compulsory for compromise minimization.

5. Hybrid Organic Inorganic Polymeric Composites Made from PES Fibrous Matrices Covered/Finished with Bulk/Micro Scaled/Powdered ZnO as Candidates to Anti-UV and High Hydrophobic Character

In order to tune the wetting behavior of 100% PES fibrous matrice by padding, functionalization with ZnO was carried out.. Two dispersion media were used for the ZnO commercially available powder, namely water and methanol, with various concentrations. The best coatings for PES fibrous matrices surface were obtained for 3% and 5% in both cases. EDX analysis performed on prepared samples confirmed the presence of ZnO on the fiber surfaces with variable coverage. Surface wettability study showed that 3% ZnO concentration in any of the preparation conditions corresponds to a maximum of the CA of about 120°C and best surface hydrophobicity. The use of methanol, as dispersion medium, leads to slightly higher values of CA with the best hydrophobicity for 48 -hour-prepared samples. Moreover, thermal stability of prepared samples was investigated by TGA. Results show that ZnO-PES hybrids reduced thermal stability compared with uncoated PES samples. ZnO presence on fibrous matrices leads to a dramatic decrease in decomposition temperature, the sample prepared from 5% ZnO in dispersion beginning decomposition at 150 °C. This may advance the observation that the coating finishing with ZnO does not improve the thermal stability of PES fibrous matrices, but the opposite.

The results of this study might offer an important contribution to stable superhydrophobic ZnO modified surfaces belonging to polymeric fibrous matrices, by manufacturing in a simple and scalable manner, which allow surface tailoring, and water barrier attributes, with direct impact on comfort performance.

In order to obtain polymeric fibrous matrices with combined properties, such as UV irradiation protection and stability, water and vapor repellence, an inorganic–organic material, based on ZnO microparticles impregnation in a polymeric network - simple and easier to implement in practice – was used. The optimum level of impregnation of the fibrous matrices is established to 3–5% ZnO, yielding stable properties, without overloading the fabric. The good homogeneity and dispersion of ZnO microparticles into the polymeric fibrous matrix ensure better UV absorbing and dissipating of the surface charge, as well as higher hydrophobic character. The ZnO microparticles impregnation method keeps the mechanical properties of the polyester fibrous matrix unchanged offering enhanced performance under environmental conditions, comfort and time stability of the garments.

The textural information used in this research can be used efficiently for precise structural determination of new synthesized compounds, made of ZnO structures on fibrous supports, in order to achieve enhanced attributes of hydrophobicity and self-cleaning.

In: Novel Approach of Added-Value Zinc Oxide ... ISBN: 978-1-63483-432-2
Editors: N. Vrinceanu and P. Postolache © 2015 Nova Science Publishers, Inc.

Chapter 4

SCIENTIFIC, PROFESSIONAL, AND ACADEMIC FUTURE DEVELOPMENT-PLANS

Narcisa Vrinceanu*[*], *PhD, and Alina Gligor, PhD
"Lucian Blaga" University of Sibiu, Romania

This part is dedicated to the intention of career development in the area of renewable polymeric matrices, a topic in which the editor aims at conducting, developing and forming a research-innovation direction for the doctoral studies school.

There are two directions envisaging (a) the synthesis of cellulose-derived materials, and (b) their application as new friction materials as an alternative to brake pads made of asbestos fibres.

The evolution plans aspire to encourage theme-based cooperation between academic environment and industrial media. Thus, the intention is to work on new product development and formulation in a program with strong industry engagement. Specifically, the proposed plan will focus closely on applied research and the expansion of the transfer of technology and knowledge in the automotive industries and beyond. Subsequently we propose an interdisciplinary research and development program for material sciences and technology development, meaning:

[*] Corresponding author Address. Email: vrinceanu.narcisai@ulbsibiu.ro.

1. Development of new improved ecological friction materials;
2. Collecting/entrapping/ destroying the brake wear particles as they are produced by photocatalytic methods (filters).

PROBLEMS

Nowadays, one of the most important topics in the context of environmental and health hazard research is airborne particles. The exposure to traffic-related particulate matter (PM) plays an important role in the development of adverse health effects. Traffic-generated particles are associated with cardiovascular and pulmonary effects, as shown by both epidemiological and toxicological studies [103, 87, 154, 144, 139, 140, 45, and 125]. Some of these studies suggest that brake wear particles might contribute to these adverse effects. Specific health concerns are related to the allergenicty of tire components such as polycyclic aromatic hydrocarbons (PAHs) and latex, and the heavy metals used in brake linings. All of these compounds have toxic, mutagen attributes [82], but above all, there are the respiratory sensitizers, inducing irreversible allergic reactions in the respiratory system.

Whilst exhaust gases in the road transport are already a concern for the European directives, there are currently no legal requirements for the control of non-exhaust particle emissions. Consequently, some of the approaches relating to non-exhaust particulate matter were highlighted [17]. Researches reveal that the traffic-related emissions are major sources of suspended PM in the urban areas [28, 85, 86]. Nevertheless, it can be claimed undoubtedly, that data relating to the physical and chemical properties, emission rates, and health consequences of non-exhaust particles derived from specific wear brake are uncertain/far from comprehensive.

Consequently, consistent knowledge concerning the effectiveness of the emission strategies is mandatory, in order to develop practical strategies for reducing the pollutant concentrations.

The research plan proposes two novel directions tightly bound to my double expertise, *knowledge of the cellulosic polymeric supports treatment against ageing process*, gained during the doctoral studies, and a *higher comprehension of synthesis and characterization of nanoporous materials used in photocatalysis of organic pollutants* derived from textile industry wastewaters, during my post-doctoral activity. These two directions are the following: (i) development of new ecological friction materials; (ii) collecting the brake particles as they are produced.

RATIONALE OF THE RESEARCH-INNOVATION PLAN

The development of new friction materials as an alternative to brake pads made of asbestos fibres sprang from the context of environmental and health hazard within global climate change. During braking, asbestos could generate wear particles of different chemistries, which are released into the environment, with a different potential of toxicity and mutagenicty.

Moreover, there are various problems associated with asbestos due to its adverse effect on problem of life cycle assessment and waste management at the end of its service life. Since asbestos was banned, the number of its substitutes has been tested in formulations of the friction composites. The typical example is aramid fibers for the non-asbestos organic low metallic (or nonmetallic) friction composites. Nevertheless, aramid fibers are still of synthetic origin, being prepared by means of chemical synthesis, which could be hardly regarded as an environmentally friendly process.

At this point, the idea of asbestos replacement with friction materials having at least the same friction performance, given by its main attributes thermal, electrical, and sound insulation, and inflammability, appeared. There were trials to combine asbestos fibers with cement or to be woven into fabric or mats/ sheets.

Remarkable research is being done worldwide for the development of biodegradable polymers, for various engineering applications [98]. The present plan proposes that plant fibers, such as sisal, jute, hemp, flax, should be used as reinforcement for different polymers, in place of synthetic fibers (glass, Kevlar, carbon, etc.). Being renewable and biodegradable materials, cellulosic fibers are one of the highest valued alternatives to be used as components in different kinds of composites applications (e.g. kenaf fibers and betelnut fibers) [118, 102, and 14]. Shalwan and Yousif summarized the literature data and worked out a comprehensive review of the utilization of natural fibers as components in polymer matrix composites, with a special focus on their tribological behavior [158].

The main demands, which these new friction composites for brake linings, proposed by this career plan, should fulfill, are the following:

- Acceptable values of the friction coefficient;
- Stability at higher temperatures.

The most extremely serious aspect is the one regarding the thermal stability of the plant fibers. In terms of thermal stability, in comparison with the most valued aramid fibers, whose decomposition interval ranges from 400 to 450°C, these temperatures are lower.

Since, during intense breaking, the temperature at the friction contact surface can locally exceed several hundred °C, the question is whether the natural plant fibers can be used for application at such higher temperatures.

The research plan stresses the idea of using three main approaches to increase plant fibers in thermal stability:

- Increasing the content of cellulose, by means removing the hemicellulose and lignin content by an alkaline procedure;
- The improvement of thermal stability of the new obtained friction polymeric composites using montmorillonite (MMT) emulsion, which is a type of inorganic natural clay containing SiO_4 tetrahedral sheets arranged into a two-dimensional network structure which might provide thermal resistance, based on the knowledge gain of scientific activity during the post-doc research [183].
- The direct growth of pure and doped nanostructured ZnO coating onto the above mentioned fibrous polymeric matrix as active photocatalyst material in air purification reactors. In other words, the research plan proposes the creation of some polymeric filters, whose place remains an important question.

Based on the work experience in the metal oxide and other photoactive materials for coatings area, a novel approach to the direct growth of pure and doped nanostructured ZnO coating onto different kinds of polymeric supports as active photocatalyst material in air purification reactors, is proposed.

By controlling the growth kinetics, it is possible to change the growth behavior of ZnO nanostructures. Growth of ZnO nanostructures can be achieved in a cheaper way at low temperature using chemical growth techniques such as aqueous chemical growth, nonaqueous solution growth, sol gel and spray deposition. All these techniques allow the use of various polymeric supports for nanostructured ZnO growth having the advantage of a much larger available surface than usual, flat substrates leading to an increased surface in volume ratio of the active photocatalyst.

IMPACT

The expected impact of the *first research direction* is to make available new scientific knowledge regarding friction materials able to consistently reduce the emissions of brake wear particulate matter, since there is a niche in this field. The expected results will be a valuable addition to the knowledge in the field of novel materials with friction performance, because ecological friction materials are in their infancy. This research might open a broad field of technological applications, due to their cost and low environmental impact of the eco-pad, with potential commercialization in the future. Since the research plan targets air purification in fact, the achieving of the fundamental objectives will have a strong environmental, social, and economic impact.

The replacement in the model formulation of mineral fillers and fibers (e.g. glass fibers); metals, such: as copper, lead, tin; antimony trisulfide, and aramid pulp, presently used in friction materials, by renewable natural fiber-reinforced- polymeric matrix composites (NF: cotton, flax or sisal fibers) will lead to:

- Thanks to their biodegradability, the production of vehicle components with potentially complete recyclability will be sustainable (zero waste at the end of the life cycle);
- An achievable lower weight will mean a decrease of gas emission in vehicles enhancing the quality of life, due to the low density of the NF;
- More safety during fiber managing and a longer life for processing tools will be provided by the lower abrasiveness and friendly handling of natural fibers as reinforcing elements;
- Positive health effects;
- Eco-friendly, lack of toxicity;
- Low cost and weight, with potential perspective commercialization;
- Better ratio properties/weight, compared to glass fibers on the expense of lower structural properties;
- Mechanical properties identical to those of traditional reinforcements, reduced tool wear and tear, high geometric stability of the manufactured parts, good insulation characteristics.

The newly developed eco-pad will be made from non-asbestos, non-metal, and non-aramid fricton materials, which meets all the developing trends of the modern friction industry.

In terms of *the second research direction*, the expected impact is to make accessible new scientific information regarding photocatalytic materials on polymeric support, in order to capture/block/entrap the emissions of brake wear particulate matter, where this kind of knowledge is uncertain.

OBJECTIVES

A. For the New Friction Materials

The novelty/originality of the proposed research plan consists in the application of life cycle assessment (LCA) and eco-design methods to evaluate and optimize friction material formulations, through a comprehensive consideration of resources consumption and bio-toxicity of friction debris. In the above context, and given the variety and complexity of non-exhaust emission sources, the statement proposes some specific objectives, obviously derived from the fundamental objective of the research plan.

- To obtain new ecological enhanced friction composites (eco-pads). The role of plant fibers in these nonmetallic friction composites will be studied in relation to formulation, friction performance, and friction surfaces.
- To optimize of the product/process: processability and performances of the composites. This aspect requires an optimization of the matrix components and the processing procedures. It is compulsory to optimize composition (percentage of polymers and fiber; type and amount of additives and filler), procedure (mixing, extrusion; treatment offibers), and process conditions (pH, temperature).
- To apply the extension evaluation method of brake friction materials, which is an effective tool for the ranking/selection of friction materials, based on some defined criteria as follows: performance, physical properties, costs of raw materials, brake effectiveness, wear, and thermal stability.
- To create a small-scale prototype pad, integrating the best friction behavior to reduce the emission of brake wear particles up to 50%.

- To create a database for environmental impacts and bio-toxicity of friction materials, to evaluate and optimize the friction material formulations through a comprehensive consideration of resources comsumption and bio-toxicity of friction debris.

B. For the Photocatalytic Materials

The originality of this carrier plan consists in the use as growth support of polymeric materials to increase photocatalytic active material surface, the use of the simple chemical approach for direct growth of nanostructured ZnO, as well as the use of ZnO based material as photocatalyst in an innovative prototype reactor/filter.

1. To obtain nanostructured ZnO based coatings on polymeric supports with controlled thickness, very good coverage, and robustness using as direct growth method different chemical simple routes (aqueous chemical growth, nonaqueous solution growth, and sol gel) using low cost precursors, and repeatability for potential commercial applications.
2. To find the optimum material composition, structure and properties in each textile substrate and to achieve maximum surface to volume ratio for active coating, for each chemical growth method.
3. To achieve materials with high photocatalytic activity against the emissions derived from the brake wear, such as air purification systems.
4. To create a small-scale prototype reactor, integrating the best photocatalytic active ZnO based material onto polymeric support, in terms of sustainability, feasibility and cost effectiveness for future scale up.

METHODOLOGY

A. For the New Friction Materials

1. **Preparation of eco-friendly friction composites with tailored friction formulations**, given my expertise on cellulosic supports behavior:

- Cellulose based fibers in composites formulation;
- *Cotton composite*: polyester resin reinforced with montmorillonite coated cotton fibers;
- *Banana/pineapple composite*: epoxy resin reinforced with banana/pineapple fibers;

  Sample preparation:
  - Extract fibers;
  - Prepare epoxy and hardener;
  - Prepare mold;
  - Fabricate composite.
- *Jute/ramie* fibers in mixture with powdered nut shells as natural and biodegradable fillers in non-asbestos organic (NAO) friction composites (graphite will be replaced with nut shell);
- *Jute composites*: Jute/-polyester-clay (montmorillonite) composite, Jute vinyl ester composite, Jute epoxy composite, Jute/polypropylene/ polystyrene thermoplastic composite; Montmorillonite coated jute fiber; reinforced polyvinyl chloride (PVC) film composites;
- *Sisal composites*: Montmorillonite coated sisal/polyester composite, Sisal/epoxy composite, Sisal Urea/formaldehyde composite, Sisal polystyrene/polypropylene composite;
- *Flax composite*: Montmorillonite coated flax polyester composite, Flax/epoxy composite, Flax/Polystyrene composite, Flax/poly-propylene composite;
- *Hybrid composite* with different volume fraction of glass and bamboo fibers reinforced with an epoxy polymer;
- *Wool–polyester* resin composites;
- *Aramid pulp and natural fibers.*

A special attention will be given to asbestos-free brake pad using agricultural wastes as a source of raw materials (coconut shell). In terms of economical wastes, utilization is not a favorable solution, but the result might be environmentally monitoring/control.

Method of production: dry wastes, grind, make different sized sieves, and use a compression molding machine.

Methods of characterization: compressive strength test, flame resistance, water and oil (SEA 20/50) absorption, and wear rate.

Methods for the making of natural fiber composites [207]:

a. Filament winding
b. Hand Lay-up/Spray consisting of two processes:

Characterization methods of eco-friendly friction composites by analytical techniques will be combined to study the topography, microstructure, chemical composition, and thermal stability of the friction materials, including SEM – for Morphology, EDX – for elemental analysis, XRD, TGA, EMA, XRF, and profilometry.

Identifying the compounds in the raw components of brake pads by: Atomic Absorption Spectrometry (AAS), and Atomic Emission Spectrometry (AES) with Inductively Coupled Plasma, Spectrometry (ICP);

Physical properties: bulk density; water absorption properties;

*Tensile properties: t*est direction; fracture stress; fracture strain; Young's Modulus;

*Mechanical properties: t*ensile strength and elongation break;

Thermal properties by DTA and TGA analyses

Measurements of friction performance by:

- *F*riction tester according to the SAE J661 recommendation [150];
- Constant-speed friction test machine;
- Friction assessment and screening test (FAST) machine;
- Automotive single-ended full-scale brake dynamometer.
- *Abrasive wear test (Taguchi method)*

Characterization of microstructures of friction surfaces/layers of the brake friction composites:

a. Friction surface morphology by: scanning electron microscopy with energy dispersive X-ray microanalysis and electron microprobe analysis, light microscopy (LM);

Phase analysis:

b. X-ray powder diffraction methods;
c. X-ray fluorescence spectrometry;
d. Thermal gravimetric analysis.

Identification of:

(1) Dynamic behavior: the structure of friction layers changes according to the different superficial positions and across sample's thickness;
(2) Friction condition dependence: the formation of friction layers depends upon temperatures, time, and thermal history such as fade and recovery;
(3) Compositional dependence: the compositions of friction surface and bulk, differ; nevertheless, bulk composition determines the friction layers.

Evaluation of:

- Braking performance characteristics;
- Thermal Stability and wear properties;
- Braking system performance assessment (BEEP);
- Relationships between friction performance and compositions of the friction composites.

B. For the Photocatalytic Materials/Filters

1. Direct growth of pure and doped ZnO nanostructured material onto various polymeric supports, by chemical techniques (aqueous chemical growth, nonaqueous solution growth, and sol gel).

1.1. Growth method optimization with respect to the substrate. Materials will be selected as potential support for the photocatalytic ZnO layer (natural fiber polymeric support, synthetic fibrous matrices, and mineral fiber fibrous matrices) with different textures. Due to the targeted application the materials with low optical absorption in UV-Vis region of the light spectrum will be preferred.
1.2. ZnO growth parameters optimization for each support.
1.3. Mn, Ag, Cu, Ni, V, and S, with various concentrations, will be used as dopants in ZnO lattice in order to enhance its photocatalytic activity in the visible region of electromagnetic spectrum and tailor morphology.

2. Systematic study of structural, chemical, and physical properties of pure and doped ZnO nanostructured material grown on various polymeric supports, and growth optimization with respect to the substrate and dopant.

2.1. Pure and doped ZnO material characterization: study of topology of the coating involving optical and electron microscopy SEM, TEM, AFM, structure, composition and stoichiometry XRD, EDX, FT-IR, Raman, BET, optical properties using UV-VIS spectroscopy.
2.2. Coated polymeric supports characterization testing: coating quality inspection using microscopic techniques, optical properties in the case of transparent textile substrates, adherence studies, washing tests, hydrophilic behavior evaluation.

3. Evaluation of the photocatalytic activity against for each material on each substrate followed by optimization of the growth technique with respect to the photocatalytic efficiency.

1.3. Preliminary photocatalytic tests employing wear debris/brake dust from the inside of the brakes;
1.4. Photocatalytic activity against various particles generated by brake wear. The tests will employ various light sources. The tests will be done for one contaminant at a time in synthetic air. The pollutant depletion will be monitored for a reasonable time and the results will be compared to establish which pair photocatalyst-substrate has the highest activity.

4. Integration of the optimum coated polymeric material in a novel laboratory small-scale prototype of a photocatalytic reactor, which may be the subject of a patent application.

4.1. Design of small laboratory scale urban air decontamination prototype reactor to host the best-engineered photocatalytic active material;
4.2. Execution of the designed prototype reactor for air decontamination and integration of the best photocatalytic active material;
4.3. Testing the prototype reactor for air decontamination and optimization with respect to application and aesthetical factors.

REFERENCES

[1] O.V. Abramov, A. Gedanken, Y. Koltypin, N. Perkas, I. Perelshtein, E. Joyce, T.J. Mason, Pilot scale sonochemical coating of nanoparticles onto textiles to produce biocidal fabrics, *Surface and Coatings Technology* 204 (2009) 718–722.

[2] Aigbodion, S., Akadike, U., Hassan, S.B., Asuke, F., Agunsoye, J.O. Development of Asbestos -Free Brake Pad Using Bagasse, *Tribology in industry* 2010; 32(1).

[3] J. Aizenberg, S. Albeck, S. Weiner, L. Addadi, *J. Am. Chem. Soc.* 1993, *115,* 11691-11697.

[4] Allen NS, Edge M, Corrales T, Childs A, Liauw CM, Catalina F, et al. Ageing and stabilization of filled polymers: an overview. *Polym Degrad Stab.* 1998;61:183-99

[5] Anderson, R.B. Modifications of the Brunauer, Emmett and Teller Equation, *J Am Chem Soc,* 68(4): 686-691 (1946).

[6] S. Anitha, B. Brabu, D. John Thiruvadigal, C. Gopalakrishnan, T.S. Natarajan, Optical, bactericidal and water repellent properties of electrospun nano-composite membranes of cellulose acetate and ZnO, *Carbohydrate Polymers* 87 (2012) 1065–1072.

[7] G. Applerot, N. Perkas, G. Amirian, O. Girshevitz, A. Gedanken, Coating of glass with ZnO via ultrasonic irradiation and a study of its antibacterial properties, *Applied Surface Science* 256 (2009) S3–S8.

[8] M.H. Asif, F. Elinder, M. Willander, Electrochemical biosensors based on ZnO nanostructures to measure intracellular metal ions and glucose, Journal of Analytical and Bioanalytical Techniques S7 (2011) 1–9.

[9] E.S. Ates, H.E. Unalan, Zinc nanowire enhanced multifunctional coatings for cotton fabrics, *Thin Solid Films* 520 (2012) 4658–4661.

[10] Becheri, M. Dürr, P. Lo Nostro, P. Baglioni, Synthesis and characterization of zinc oxide nanoparticles: application to textile as UV-absorbers, *Journal of Nanoparticle Research* 10 (2008) 679–689.
[11] Bender ML, Komiyama M. *Cyclodextrin Chemistry*, Springer, 1978, Berlin.
[12] Bergamasco RC, Zanin GM, De Morales FF *J. Inclusion Phenom Macrocyclic Chem*. 2007; 57, 157.
[13] J. Biscan J., *Croatica Chemica Acta*, CCACAA 80(3-4), (2007) 357
[14] Bledzki, A.K., Gassan, J. Composite reinforced with cellulose based fibers. *Prog Polym Sci* 1999;24:221.
[15] De Boer, J.H. *The Dynamical Character of Adsorption*, 2nd ed., Clarendon Press, Oxford, 200-219 (1968).
[16] Borcia, G. Borcia, N. Dumitrascu, Surface treatment of polymers by plasma and UV radiation, *Romanian Journal of Physics* 56 (2011) 224–232.
[17] Boulter, P.G. (2005). A review of emission factors for road vehicle non-exhaust particulate matter. TRL Report PPR065 TRL Limited, Wokingham.
[18] Brayner R, Ferrari-Iliou R, Brivois N, et al. Toxicological impact studies based on Escherichia coli bacteria in ultrafine ZnO nanoparticles colloidal medium. *Nano Lett* 2006; 6: 866.
[19] G. Broasca, C. Campagne, D. Farima, M. Ciocoiu, M. Iorgoaea, Vrinceanu N., *The International Scientific Conference UGALMAT 2011*, Advanced materials and technologies, Galati, Romania, (2011) 175.
[20] Broasca G., Suchea M.P., Ciocoiu M., Farima D., Narcisa Vrinceanu, Campagne C., Rimbu C., Iorgoaea M.: „Novel Approach Regarding Zeta Potential Variation As A Consequence Of a Woven Fabrics Antibacterial Finishing Treatment*", Tekstil ve Konfeksiyon" (Journal of Textile and Apparel)*, 23(1), 2013, 43-48, ISSN 1300-3356.
[21] Broasca G., Borcia G., Dumitrascu G., Narcisa Vrinceanu: „Characterization of ZnO coated polyester fabrics for UV protection", *Applied Surface Science*, 279, 2013, pp. 272– 278, ISSN 0169-4332.
[22] Brunauer, S., Deming, L.S., Deming, W.E., Teller, E. On a Theory of the van der Waals adsorption of gases, *J Am Chem Soc*, 62(7): 1723-1732 (1940).
[23] Buksek H., Luxbacher T. and Petrinic I., 2010, „Zeta Potential Determination of Polymeric Materials Using Two Differently Designed Measuring Cells of an Electrokinetic Analyzer", *Acta Chim. Slov.*, Vol:57, pp. 700–706.

[24] Cao, G. *Nanostructures & Nanomaterials: Synthesis, Properties & Applications*, Imperial College Press, 2008. – model de carte.

[25] Chen B, Peng X, Wang J, Sun S. Comp. Mat. Sci. 2008; 4:201.

[26] Chen CG, Khobaib M. Curliss D. Epoxy layered silicate nanocomposites. *Prog Org Coat.* 2003;47:376-383.

[27] Chibowski E and Mittal KL. Contact angle, wettability and adhesion. Vol. 2, Utrecht: VSP, 2002, p.265.

[28] Chow, J.C., Watson, J.G., Egami, R.T., Franzier, C.A, Lu, Z., Goodrich, A. Bird, A. Evaluation of regenerative-air vacuum street sweeping on geological contributions to PM_{10}. Journal *of the Air and Waste Management Association,* 1990;40 (8):1134-1142.

[29] Christopher PW, Bruce HJ. *Sedimentary Research* 1988; 58.

[30] Corrales T, Catalina F, Peinado C, Allen NS, Fontan E. Photooxidative and thermal degradation of polyethylenes interrelationship by chemiluminescence, thermal gravimetric analysis and FT-IR data. J *Photochem Photobiol.* A 2002;147:213-224.

[31] N. Didane, S. Giraud, E. Devaux. Thermal and fire resistance of fibrous materials made by PET containing flame retardant agents. *Polymer Degradation and Stability*, 97 1083-1089 (2012).

[32] Drexler, E. K. *Engines of Creation: The Coming Era of Nanotechnology;* Anchor Books: New York, US, 1986.

[33] N. Dura'n, D. Marcato Purity, G.I.H. De Souza, O.L. Alves, E. Esposito, Antibacterial effect of silver nanoparticles produced by fungal process on textile fabrics and their effluent treatment, *J Biomedical Nanotechnology* 3 (2007) 203–208.

[34] M. Eita, L. Wagberg, M. Muhammed, Spin-assisted multilayers of poly(methyl methacrylate) and zinc oxide quantum dots for ultraviolet-blocking applications, *ACS Applied Materials & Interfaces* 4 (2012) 2920–2925.

[35] El-Rafie, H.M.; El-Rafie, M.H.; Zahran, M.K. (2013). Green synthesis of silver anoparticles using extracted from marine macro algae. *Carbohydrate Polymers 96*, 403–410.

[36] Faal N, Farzaneh F. *App. Surf. Sci.* 2006; 17(3):231-234.

[37] Fan Q, John J, Ugbolue SC, Wilson AR, Dar YS, Yang Y. Nanoclay-modified polypropylene dyeable with acid and disperse dyes. *AATCC Rev.* 2003; 3(6):25.

[38] Richard P. Feynman, Robert B. Leighton, and Matthew Sands, *The Feynman Lectures on Physics*, Addison-Wesley, Reading, MA, Vol. 1, 1964, p. 12-2.

[39] Fujihara S, Sasaki C, Kimura T. *Appl. Surf. Sci.* 2001; 180:34.

[40] Furuta T, Kusuya Y, Neoh TL, Rehmann L, Beak SH, Yoshii H. *J. Inclusion. Phenom. Macrocyclic Chem* 2006; 56:107.

[41] Gardner TJ, Messing GJ. *Thermochim. Acta* 1984; 78:17.

[42] Garvey CJ, Parker IH, Simon GP. On the interpretation of X-ray diffraction powder patterns in terms of the nanostructure of cellulose fibres. *Macromol Chem Phys*. 2005;206:1568–75.

[43] Gawas UB, Mojumdar SC, Verenkar VMS. *J Therm Anal Calorim.* 2009;96:49-52.

[44] Gawas UB, Verenkar VMS, Mojumdar SC. Synthesis and characterization of $Co_{0.8}Zn_{0.2}Fe_2O_4$ nanoparticles. *J Therm Anal Calorim*. 2011;104:879-883.

[45] Gerlofs-Nijland, ME, Dormans JAMA, Bloemen HJT, Leseman DLAC, Boere AJF, Kelly FJ, Mudway IS, Jimenez AA, Donaldson K, Guastadisegni C, Janssen NAH, Brunekreef B, Sandström T, van Bree L, Cassee FR. *Toxicity of coarse and fine particulate matter from sites with contrasting traffic profiles. Inhal Toxicol* 2007;19: 1055-1069.

[46] Gowri, V.S., Saxena, M. Protection of Bamboo Surfaces by CNSL Based Coatings, *J Chem Technol 14:145-149 (1997).*

[47] Gonsalves LR, Mojumdar SC, Verenkar VMS. Synthesis and characterization of $Co_{0.8}Zn_{0.2}Fe_2O_4$ nanoparticles. *J Therm Anal Calorim*. 2011;104:869-873.

[48] Grigoriu A.M., Cercetări în domeniul compusilor de incluziune ai ciclodextrinelor si al derivatilor acestora cu aplicatii în industria textilă. Ph. D. Diss., Iasi (2009).

[49] Grigoriu AM, Luca C. Noi aplicaţii ale ciclodextrinelor în industria textilă PERFORMANTICA Publishing House 2010, Iasi.

[50] Grigoriu AM, Luca C, Vrinceanu N, Ciolacu F. Inclusion compounds of monochlorotriazinyl-β-cyclodextrin grafted on a paper support, *Cellul. Chem. Technol*. 2011;45(3-4):177-183.

[51] Gu, F., Wang, S.F., Lu, M.K., Zhou, G.J., Xu, D., Yuan, D.R. Stucture evaluation and highly enhanced luminescence of Dy3+-doped ZnO nanocrystals by Li+ doping via combustion method. *Langmuir*, 20: 3528 (2004).

[52] Guggenheim, E. A. *Application of Statistical Mechanics*, Clarendon Press, Oxford, 186-206 (1966).

[53] Gui XC, Wei JQ, Wang KL, et al. Carbon Nanotube Sponges. Adv Mat 2010; 22: 617.

[54] Guo L, Yang S, Yang C, Yu P, Wang J, Ge W, Wong GKL. *Chem. Mater*. 2000; 12: 2268–2274.

[55] Guo L., Campagne C., Perwuelz A. and Leroux F.,2009,'' Zeta Potential and Surface Physico-chemical Properties of Atmospheric Air-plasma-treated Polyester Fabrics'', *Textile Research Journal,* Vol 79(15), 1371–1377.

[56] K.L. Hatch, U. Osterwalder, Garments as solar ultraviolet radiation screening materials, *Dermatologic Clinics* 24 (2006) 85–100.

[57] He Y, Yang B, Cheng G. Catal. *Today* 2004; 98: 595.

[58] He, J.H., Kunitake, T., Nakao, A., Facile in situ synthesis of noble metal nanoparticles in porous cellulose fibers, *Chem. Mater*. 15, 4401–4406 (2003).

[59] He, J.H., Kunitake, T., Nakao, A., Facile fabrication of composites of platinum nanoparticles and amorphous carbon films by catalyzed carbonization of cellulose fibers, *Chem. Commun*. 4, 410–411 (2004).

[60] Hingorani S, Pillai V, Kumar P, Multani MS, Shah DO. Microemulsion mediated synthesis of zinc-oxide nanoparticles for varistor studies. *Mater Res Bull*. 1993;28:1303–10.

[61] Huang MH,. Wu YY, Feick HN, Tran N, Weber E, Yang PD. *Adv. Mater*. 2001; 13:113.

[62] Huang Z, Zheng X, Yan D, et al. Toxicological effect of ZnO nanoparticles based on bacteria. *Langmuir* 2008; 24: 4140.

[63] Hughes, L. J., Britt, G.E. Compatibility studies on polyacrylate and polymethacrylate systems. *J. Appl. Polym. Sci.,* 5:337 (1961).

[64] D.R.J. Hunter, Zeta Potential in Colloid Science: Principles and Applications, Academic Press, London, 1981.

[65] J.A. Ibanez, J. Forte, A. Hernandez and F. Tejerina, *J. Membrane Sci.*, 36 (1988) 45.

[66] J. Israelachvili, Intermolecular & Surface Forces, second ed., Academic Press Ltd., London, 1992.

[67] H.J. Jacobasch, G. Baubock and J. Schurz, *Colloid and Polymer Sci.*, 263 (1985) 3.

[68] H.J. Jacobasch and J. Schurz, *Progr. Coll. Polym. Sci.*, 77 (1988) 40.

[69] Jayanta Kumar Behera. Synthesis and Characterization of Nano-particles.M. Tech Thesis, NIT.

[70] Rourkela.

[71] Jeevani, J. (2011). Nanomedicine & Nanotechnolology. http://omicsonline.org/nanotextiles-a-broader-perspective-2157-7439.1000124.pdf

[72] Jiadao W, Ying Y and Darong C. Research progress on the ultra hydrophobic surface topography effect. *Chinese Sci Bull* 2006; 51: 2297.

[73] Jiang, E. Newton, C.W.M. Yuen, C.W. Kan, Chemical silver plating and its application to textile fabric design, *Journal of Applied Polymer Science* 96 (2005) 919–926.

[74] Jiang, L., Yao, X., Li, H. X., Fu, Y.Y., Chen, L., Meng, Q., Hu, W.P., Jiang, L. All-Carbon ElectronicDevices Fabricated by Directly Grown Single-Walled Single-Walled Carbon Nanotubes on Reduced Graphene Oxide Electrodes. *Adv. Mater.* 22:376 (2010).

[75] S.Q. Jiang, E. Newton, C.W.M. Yuen, C.W. Kan, Chemical silver plating and its application to textile fabric design, *J Appl Polym Sc*i 96 (2005) 919–926.

[76] Jiang ZH, Zhong Y, So CS,. Hse CY. Rapid prediction of wood crystallinity in Pinus elliotii plantation wood. *J Wood Sci.* 2007;53, 449.

[77] Jimenez AA, Donaldson K, Guastadisegni C, Janssen NAH, Brunekreef B, Sandström T, van Bree L, Cassee FR. Toxicity of coarse and fine particulate matter from sites with contrasting traffic profiles. *Inhal Toxicol* 2007;19: 1055-1069.

[78] Jiugao, Yu, Jingwen, Yang, Baoxiang, Liu, Xiaofei, Ma, Preparation and characterization of glycerol plasticized - pea starch/ ZnO–carboxy-methylcellulose sodium nanocomposites, *Bioresource Technology* 100, 2832–2841 (2009).

[79] P. Jongnavakit, P. Amornpitoksuk, S. Suwanboon, T. Ratana, Surface and photocatalytic properties of ZnO thin film prepared by sol–gel method, *Thin Solid Films* 520 (2012) 5561–5567.

[80] Joshi, M. and Bhattacharyya, A. (2011). Nanotechnology—a new route to high-performance functional textiles. *Textile Progress 43, 155–233.*

[81] S. Kathirvelu, L. D'Souza, B. Dhurai, UV protection of textiles using ZnO nanoparticles, *Indian Journal of Fibre & Textile Research* 34 (2009) 267–273.

[82] Kekkonen, V., Hakola, A., Kajava, T., Sahramo, E., Malm, J., Karppinen, M., Ras, R. H. A., *Appl. Phys. Lett.* 97:044102 (2010).

[83] Kelly, F. J., & Fussell, J. C. Air pollution and airway disease. *Clinical & Experimental Allergy*, 2011;41(8): 1-13. doi:10.1111/j.1365-2222.2011.03776.x

[84] M.G.A. Khedr, S.M. Abdel Haleem and A. Baraka, *J. Electroanal. Chem.*, 184 (1985).

[85] D.B. Kittelson, Engines and nanoparticles: a review, *J Aerosol Science* 29 (1998) 575–588.

[86] Kuhns, H., Etyemezian, V., Landwehr, D., MacDougall, C., Pitchford, M., Green, M. (2001). *Testing reentrained aerosol kinetic emission from roads (TRAKER): a new approach to infer silt loadings on roadways.* Atmospheric Environment 2001;35:2815-2825.

[87] Kuhns, H., Etyemezian, V., Green, M., Hendrickson K., McGown, M., Bartond, K., Pitchford, M. *Vehicle-based road dust emission measurement- Part II: Effect of precipitation, wintertime road sanding, and street sweepers on inferred PM_{10} emission potentials from paved and unpaved roads. Atmospheric Environment*, 2009;(37): 4573-4582.

[88] Kukutschova, J., Roubicˇek, V., Malachova, K., Pavlicˇkova, Z., Holuša, R., Kubachkova,J. *Wear mechanism in automotive brake materials, wear debris and its potential environmental impact.* Wear 2009;267:807–17.

[89] Kwon YJ, Kim KH, Lim CS. Shim KB. *J. Ceramic Proc. Res.* 2002; 3:146–149.

[90] Lee HJ, Yeo SY, Jeong SH. Antibacterial Effect of Nanosized Silver Colloidal Solution on Textile Fabrics. *J Mater Sci.* 2003;38:2199–2204.

[91] Leroy P., Tournassat C., Bizi M., 2011, „The influence of surface conductivity on the apparent zeta potential of TiO2 nanoparticles'', *Journal of Colloid and Interface Science*, Vol:356(2), pp: 442-453.

[92] J. Li, Y.X. Yang, F. Zha, Z.Q. Lei, Facile fabrication of superhydrophobic ZnO surfaces from high to low water adhesion, *Materials Letters* 75 (2012) 71–73.

[93] Li Y, Duan G, Cao B, et al. Superhydrophobicity of 2D ZnO ordered pore arrays formed by solution-dipping template method. *Colloid Interf Sci* 2005; 287: 634.

[94] Y.Q. Li, S.Y. Fua, Y.W. Mai, Preparation and characterization of transparent ZnO/epoxy nanocomposites with high-UV shielding efficiency, *Polymer* 47 (2006) 2127–2132.

[95] J.P. Liu, S.C. Qu, X.B. Zeng, Y. Xu, X.F. Gou, Z.J. Wang, H.Y. Zhou, Z.G. Wang, Fabrication of ZnO and its enhancement of charge injection and transport in hybrid organic/inorganic light emitting devices, *Applied Surface Science* 253 (2007) 7506–7509.

[96] Liufu S, Xiao H, Li. *Powder Technol.* 2004; 145:20–24.

[97] Lorne B., Perrier F. and Avouac J. P., 1999, "Streaming potential measurements'', *J Geophysical Research*, Vol: 104(17): 17857–17877.

[98] H.F. Lu, B. Fei, J.H. Xin, R.H. Wang, L. Li, Fabrication of UV-blocking nanohybrid coating via miniemulsion polymerization, *J Colloid and Interface Science* 300 (2006) 111–116.

[99] Lu L, Farris T N, Chandrasekar S. 1992. Sliding Microindentation Wear Particles: Spheres in Grinding Swarf. WEAR PARTICLES: From the Cradle to the Grave, Ed. D. Dowson, C.M. Taylor, T.H.C. Childs, M. Godet, and G. Dalmaz, Elsevier, 257-263.

[100] Luca C, Grigoriu AM (2006) *Rev. Chim.* 2006; 57: 248.

[101] Ma, X.F., Chang, P.R., Yang, J.W., Yu, J.G.,. Preparation and properties of glycerol plasticized-pea starch/zinc oxide–starch bionanocomposites, *Carbohydr. Polym.* 75, 472–478 (2009).

[102] Mani G, Fan Q, Ugbolue SC, Eiff IM. Effect of nanoparticle size and its distribution on the dyeability of polypropylene. *AATCC Rev.* 2003; 3(1), 22.

[103] La Mantia, F.P., Morreale, M. *Green composites: A brief review. Composites* Part A 2011;42:579–88.

[104] Maynard, A. D., Aitken, R. J. 2007 *Assessing exposure to airborne nanomaterials: current abilities and future requirements*, Nanotoxicology 2007;1:26–41.

[105] J.M. Menter, K.L. Hatch, Clothing as solar radiation protection, in: P. Elsner, K. Hatch, W. Wigger (Eds.), *Textiles and the Skin, Current Problems in Dermatology,* vol. 31, Karger, Basel, 2003, pp. 50–63.

[106] Meuler, A. J., Smith, J. D., Varanasi, K. K., Mabry, J. M., McKinley, G. H., Cohen, R.E. Relationships between Water Wettability and Ice Adhesion. ACS Appl. Mater. Interfaces, 2:3100 (2010).

[107] H.F. Moafi, A.F. Shojaie, M.A. Zanjanchi, Photocatalytic self-cleaning properties of cellulosic fibers modified by nano-sized zinc oxide, Thin Solid Films 519 (2011) 3641–3646.

[108] Mocanu AM, Odochian L, Apostolescu N, TG-FT-IR study on thermal degradation in air of some new diazoaminoderivatives. *J Thermal Anal Calorim.* 2010;100 (2):615-622.

[109] VJ Mohanraj and Y Chen, *Nanoparticles A Review Tropical Journal of Pharmaceutical Research*, June 2006.

[110] Mojumdar SC, Moresoli C, Simon LC. Legge RL. Edible wheat gluten (WG) protein films: Preparation, thermal, mechanical and spectral properties. *J Therm Anal Calorim.* 2011; 104:929-936.

[111] Mucalo, M.R., Bullen, C.R., Arabinogalactan from the Western larch tree: a new, purified and highly water-soluble polysaccharide-based protecting agent for maintaining precious metal nanoparticles in colloidal suspension, *J. Mater. Sci.* 37, 493–504 (2002).

[112] Muresan R., Controlul prin Potentialul Zeta al unor operatii tehnologice de tratare si finisare la matcrialclc fibroase in vederea optimizarii acestora, PhD Thesis, Iasi, 1997.
[113] Mwalkambo LY, Ansell MP. Chemical modification of hemp, sisal, jute, and kapok fibers by alkalization. *J Appl Polym Sci.* 2002;84:2222.
[114] Nevell TP, Zeronian S H. *Cellulose Chemistry and Its Applications* M. Chichester: Ellis Horwood Ltd., 1985;423−454.
[115] Niu, J. J., Wang, J.N. A novel self-cleaning coating with silicon carbide nanowires. *J. Phys. Chem.* B 113:2909 (2009).
[116] Norton DP, Heo YW, Ivill MP, et al. ZnO: growth, doping & processing. Mat Today 2004; 7: 34; Becheri A, Durr M, Nostro P Lo, et al. Synthesis and characterization of zinc oxide nanoparticles: application to textiles as UV-absorbers. *Res* 2008; 10: 679.
[117] M. Nystrom, M. Lindstrom and E. Matthiasson, *Colloids Surfaces*, 36 (1989) 297.
[118] Ogata N, Ogawa T, Ida T, Yanagawa T, Ogihara T, Yamashita A. Structure and thermal/mechanical properties of drawn Nylon-6 hybrid. Sen'i Gakkaishi. 1995;51(9): 439..
[119] Oudshoorn, K. *Biogas from sisal waste – a new opportunity for the sisal industry in Tanzania.* Energy Sustain Dev, 1995;2(4):46–49.
[120] Park S, Baker JO, Himmel ME, Parilla PA and Johnson DK. Cellulose crystallinity index: measurement techniques and their impact on interpreting cellulase performance. *Biotechnol Biofuels* 2010; 3: 10.
[121] Parikh, D. V.; Fink, T.; Rajasekharan, K.; Sachinvala, N. D.; Sawhney, A. P. S.; Calamari, T. A.; Parikh, A.D. (2005). Antimicrobial silver/sodium carboxymethyl cotton dressings for burn wounds. *Textile Research Journal 75*, 134–138.
[122] Patra, S., Sarkar, S., Bera, S., Paul, K. G. K., Ghosh, R., *J. Phys. D Appl. Phys.* 42:075301 (2009).
[123] Patra, S., Sarkar, S., Bera, S., Paul, K. G. K., Ghosh, R., *J. Appl. Phys.* 108: 083507 (2010).
[124] Patscheider J, Zehnder T, Diserens M. Structure-performance relations in nanocomposite coatings. *Surf Coat Technol.* 2001;146-147:201-208.
[125] Patterson A. The Scherrer formula for X-ray particle size determination. *Phys Rev.* 1939;56(10):978–982.
[126] Peters, A., Heier, M., Trentinaglia, I., Hormann, A., Wichmann, H.E., Lowel, H. *Exposure to traffic and the onset of myocardial infarction.* N Engl *J Med.*2004;351:1721–1730.

[127] G. Placinta, F. Arefi-Khonsari, M. Gheorghiu, J. Amouroux, G. Popa, Surface properties and the stability of poly(ethylene terephtalate) (PET) films treated in plasmas of He–O2 mixtures, *J App Polym Sci* 66 (1997) 1367–1376.

[128] Popa A, Luca C, Mihailă G. Inclusion Compounds, Scientific Publishing House 1992.

[129] Popescu AC, Duta L, Dorcioman G, et al. Radical modification of the wetting behavior of textiles coated with ZnO thin films and nanoparticles when changing the ambient pressure in the pulsed laser deposition process. *Appl Phys* 2011; 110: 06432.

[130] Popovici, E.; Dvininov, E. *Nanostructured materials. Present and future*; *Nanoparticles*; Demiurg Publishing House: Iasi, RO, 2007; Vol.1.

[131] Popovici, E.; Seftel, E. *Nanostructured materials. Present and future*; *Anionic Clays*; Demiurg Publishing House; Iasi, RO, 2007; Vol. 2. Model de carte.

[132] Popovici, E.; Ignat Rudei, M. *Nanostructured materials. Present and future*. *Nanotubes*; Demiurg Publishing House, Iasi, RO, 2009; Vol. 4.

[133] Prepelita I.R., Narcisa Vrinceanu, Chirita R., Dobrin R.P., Alexinschi O.E., Brinza F., Hagiu B.A., Suchea P.M.: „New approach on the role of a polymer matrix in the strong texturation of ZnO nanopowder", *Revue Roumaine de Chimie*, 58(11-12), ISSN: 0035-3930, 871-878, 2013.

[134] *Principles Of Nanotechnology* - Molecular-Based Study of Condensed Matter in Small Systems World Scientific Publishing Co. Pte. Ltd http://www.worldscibooks.com/nanosci/5749.html

[135] Radhakrishnan, T., Georges, M.K., Nair, P.S., 2007. Study of sago starch–CdS nanocomposite films: Fabrication, structure, optical and thermal properties, J. *Nanosci. Nanotechn*ol. 7, 986–993 (2007).

[136] Rajendran R, Balakumar C, Mohammed Ahammed HA, et al. Use of zinc oxide nano particles for production ofantimicrobial textiles. *Int Eng Sci* 2010; 2: 202.

[137] Raveendran, P., Fu, J., Wallen, S.L., Completely ‘‘green" synthesis and stabilization of metal nanoparticles, *J. Am. Chem. Soc.* 125, 13940–13941 (2003).

[138] Reuscher H., Hinsenkorn R. BETA W7 MCT-new ways in surface modification. *Journal of inclusion phenomena and macrocyclic chemistry*, 25, 191–196 (1996).

[139] Revola JF, Dietricha NDD, Goring AI. Effect of mercerization on the crystallite size and crystallinity index in cellulose from different sources. *Can J Chem.* 1987;65:1724.

[140] Riediker, M., Cascio, W.E., Griggs, T.R., Herbst, M.C., Bromberg, P.A., Neas, L., Williams, R.W., DevlinR.B., Particulate matter exposure in cars is associated with cardiovascular effects in healthy young men. *American Journal of Respiratory and Critical Care Medicine* 2004; 169: 934-940.

[141] Riediker, M., Koren, H.S., The importance of environmental exposures to physical mental and social well-being, *International Journal of Hygiene and Environmental Health* 2004;207(3):193-201.

[142] Riediker M, Gasser M, Perrenoud A, Gehr P, Rothen-Rutishauser B. A system to test the toxicity of brake wear particle. *Am J Respir Crit Care Med.* 2007, pp. 177.

[143] Riva A, Algaba IM, Montserrat P. Action of a finishing product in the improvement of the ultraviolet protection provided by cotton fabrics. Modelisation of the effect. *Cellul.* 2006;13:697–704.

[144] Rimbu C., Narcisa Vrinceanu, Broasca G., Campagne C., Suchea M.P., Nistor A.: „Zinc oxide application in textile industry: surface tailoring and water barrier attributes as parameters with direct implication in comfort performance, *Textile Research Journal*, 83(20), 2013, pp. 2142-2151, ISSN 0040-5175.

[145] Rosca, C., Popa, M.I., Lisa, G., 2005. Interaction of chitosan with natural or synthetic anionic polyelectrolytes. 1. The chitosan-carboxymethylcellulose complex. *Carbohydr. Polym.* 62, 35–41 (2005).

[146] S.A. Ruffolo, M.F. La Russa, M. Malagodi, C.O. Rossi, A.M. Palermo, G.M. Crisci, ZnO and ZnTiO3 nanopowders for antimicrobial stone coating, *Applied Physics* A 100 (2010) 829–834.

[147] Russell, E., (2002). Nanotechnologies and the shrinking world of textiles. *Textile Horizons 9,* 7-9.

[148] Samal, S.S.; Jeyaraman, P.; Vishwakarma, V. (2010). *Journal of Minerals & Materials Characterization & Engineering 9*, 519-525.

[149] Sataev, M.S.; Koshkarbaeva, S.T.; Tleuova, A.B.; Aidarova, S.B.; Prokopovich, P.; Perni, S. (2013). *Novel process* for *coating textile materials* with*silver* to *prepare antimicrobial fabrics. Colloids and Surfaces A: Physicochemical and Engineering Aspects 442,* 146-151.

[150] Saxana, M, Gowri. V.S., *J Polym Compd* 24:428 (2003).

[151] Saxena, M., Pappu, A., Sharma, A., Haque, R., Wankhede, S. *Composite Materials from Natural Resources: Recent Trends and Future Potentials*, www.intechopen.com

[152] Saxena M, Gowri VS. Studies on Bamboo polymer composites with polyester amide polyol as interfacial agent J *Polym Compd* 2003; 24: 428–436.

[153] Schaller, D. Rogez, A. Braig, Organic vs inorganic light stabilizers for waterborne clear coats: a fair comparison, *Journal of Coatings Technology and Research* 9 (2012) 433–441.

[154] J. Scholz, G. Nocke, F. Hollstein, A. Weissbach, Investigations of fabrics coated with precious metals using magnetron sputter technique with regard to their antimicrobial properties, *Surface and Coatings Technology* 192 (2005) 252–256.

[155] Schwartz, J., Litonjua, A, Suh, H., Verrier, M., Zanobetti, A., Syring, M., et al. *Traffic related pollution and heart rate variability in a panel of elderly subjects.* Thorax. 2005;60:455–461.

[156] Shahrom BM. Synthesis and characterization of zinc oxide nanostructures 2008 PhD Thesis Maysia.

[157] Shaik, S., Kshirsagar, M., Ghanashyam Krishna, S.P. Tewari, D.D. Purkayastha, V. Madhurima, Growth of superhydrophobic zinc oxide nanowire thin films, *Materials Letters* 75 (2012) 51–53.

[158] U.P. Shaik, S. Kshirsagar, M. Ghanashyam Krishna, S.P. Tewari, D.D. Purkayastha, V. Madhurima, Growth of superhydrophobic zinc oxide nanowire thin films, *Materials Letters* 75 (2012) 51–53.

[159] Shalwan, A., Yousif, B.F. *In state of art: Mechanical and tribological behaviour of polymeric composites based on natural fibres.* Mater Des 2013;48:14–24.

[160] J. Shaw, *Electrophoresis*, Academic Press, London, 1969.

[161] Shirtcliffe, N. J., McHale, G., Newton, M.I., Perry, C.C. Intrinsically superhydrophobic organosilica sol-gel foams. *Langmuir*, 19:5626 (2003).

[162] Singhal M, Chhabra V, Kang P, Shah DO. Synthesis of ZnO nanoparticles for varistor application using Zn-substituted aerosol OT microemulsion. *Mater Res Bull.* 1997; 32:239-247.

[163] Smith BL, Schaffer TE, Viani M, Thompson JB, Frederick NA, Kind J, Belcher A, Stucky, GD, Morse DE, Hansman PK, *Nature* 1999; 399:761.

[164] Song L, Hu Y, Wang S, Fan W, Chen Z. A study on the synthesis and properties of polyurethane/clay nanocomposites. *Intern J Polym Anal Charact*. 2003;8(5):317.

[165] Stretz HA, Koo J H, Dimas VM, Zhang Y, Flame retardant properties of polycarbonate/montmorillonite clay nanocomposite blends. *Polym.* 2001;42(2): 50.

[166] Sunkyu Park, John O Baker, Michael E Himmel, Philip A Parilla and David K Johnson, RCellulose crystallinity index: measurement techniques and their impact on interpreting cellulase performance, *Biotechnology for Biofuels* 2010, 3:10.

[167] A.Sze, D. Erickson, L.Ren, and D. Li, 2003, *J. Coll. Interface Sci.*, 261, 402–410.

[168] Szejtli J. *Cyclodextrins and their Inclusion Complexes*, Akadmemiai Kiado, 1982, Budapest.

[169] Syring, M., et al. *Traffic related pollution and heart rate variability in a panel of elderly subjects.* Thorax. 2005;60:455–461

[170] Tanasa D., Narcisa Vrinceanu, Nistor A., Hristodor C.M., Popovici E., Bistricianu I.L., Brinza F., Chicet D., Coman D., Pui A., Grigoriu A.M., Broasca G.: „Zinc oxide-linen fibrous composites: morphological, structural, chemical and humidity adsorptive attribute*", Textile Research Journal*, 82(12), 2012, pp. 832-844.

[171] Tang E, Cheng G, Ma X. *Powder Technol*. 2006a; 161:209–214.

[172] Tang E, Cheng G, Ma X, Pang X, Zhao Q. *Appl. Surf. Sci.* 2006b; 252:5227–5232.

[173] Tang Z K, Wong G K L, Yu P, Kawasaki M, Ohtomo A, Koinuma H and Segawa Y., Room-temperature ultraviolet laser emission from self-assembled ZnO microcrystallite thin films. *Appl Phys Lett* 1998; 72: 3270.

[174] Taniguchi, N. On the Basic Concept of Nano-Technology. *Proceedings of the International Conference on Products Engineering Tokyo*, 1974 Part II, Japan Society of Precision Engineering.

[175] M. Tasaka, S. Tamura, N. Takemura and K. Morimoto, *J. Membrane Sci.,* 12 (1982).

[176] Taubert A, Palms D, Weiss OO, Piccini M-T, Batchelder DN. *Chem. Mater*. 2002; 14:2594.

[177] Taubert, A., Wegner, G., Formation of uniform and monodisperse zincite crystals in the presence of soluble starch. *J. Mater. Chem.* 12, 805–807 (2002).

[178] Topala, N. Dumitrascu, V. Pohoata, Influence of plasma treatments on the hemocompatibility of PET and PET + TiO2 films, *Plasma Chemistry and Plasma Processing* 28 (2008) 535–551.

[179] Tserki V, Zafeiropoulo N E, Simon F, Panayiotou C. A study of the effect of acetylation and propionylation surface treatments on natural fibres. *J Appl Sci Manuf.* 2005; 36, 1110.

[180] Ursache M., Loghin C., Muresan R., Cerempei A. and Muresan A, 2011," Investigation on the effects of antibacterial finishes on dyed cotton knitted fabrics'', *Tekstil Ve Konfeksyon*, Vol: 21.

[181] Vanheusden K, Seager CH, Warren WL,. Tallant DR,. Voigt JA. *Appl. Phys. Lett.* 1996; 68:4034

[182] Verdu J, Rychly J, Audouin L. *Polym Degrad Stab.* 2003;79:503-9.

[183] Vigneshwaran, N., Kumar, S., Kathe, A.A., Varadarajan, P.V., Prasad, V., Functional finishing of cotton fabrics using zinc oxide–soluble starch nanocomposites, *Nanotechnology* 17, 5087–5095 (2006).

[184] Vrinceanu, N., Tanasa, D., Hristodor, C.M., Brinza, F., Popovici, E., Gherca, Pui, A., Coman, D., Carsmariu, A., Bistricianu, I.L., Broasca, G. Synthesis and characterization of zinc oxide nanoparticles: application to textiles as thermal barriers, *J Therm. Anal Calorim* 2013; 111(2):1107-1119, DOI: 10.1007/s10973-012-2269-7; ISSN: 1572-8943].

[185] Xia, H.L., Honglai, D., Yarning, I., Ying, H. Analysis of modulus- strain curves of polyester yarn and cotton yarn.Chinese J. *Chem. Eng. 13:*498 (2005).

[186] H.Y. Xia, T.H. Liu, L.N. Gao, L. Yan, J.W. Wu, Development of film sensors based on ZnO nanopraticles for amine gas detection, *Applied Surface Science* 258 (2011) 254–259.

[187] Xiao-Juan J, Pascal Kamdem D. Chemical composition crystallinity and crystallite cellulose size in populus hybrids and aspen. *Cellul Chem Technol.* 2009;43(7-8):229-234.

[188] Xu T, Xie CS. Characterization of the dispersion of tetrapod-like nano-ZnO whiskers in acrylic resin and properties of the nano-composite coating system. *Prog Org Coat.* 2003;46:297-301.

[189] C.H. Xue, W. Yin, S.-T. Jia, J.-Z. Ma, UV-durable superhydrophobic textiles with UV-shielding properties by coating fibers with ZnO/SiO2core/shell particles, *Nanotechnology* 22 (2011) 415-603.

[190] Zhang J, Sun LD, Pan HY, Liao CS, Yan CH. *New J. Chem.* 2002; 26:33.

[191] Zhang, Z., Chen, H., Zhong, J., Saraf, G., Lu, Y., *J. Electron. Mater.* 36:895 (2007).

[192] Zhao N, Xie Q, Weng L, et al. Progress in Superhydrophobic Surfaces. *Macromolecules* 2005; 38: 8996;

[193] Walsh, D., Arcelli, L., Ikoma, T., Tanaka, J., Mann, S., Dextran templating for the synthesis of metallic and metal oxide sponges, *Nat. Mater.* 2, 386–390 (2003)

[194] H.H. Wang, S.J. Dong, Y. Chang, X.P. Zhou, X.B. Hu, Microstructures and photocatalytic properties of porous ZnO films synthetized by chemical bath deposition method, *Applied Surface Science* 258 (2012) 4288–4293

[195] Wang RH, Xin JH, Tao XM. *Inorg. Chem.* 2005; 44:3926–3930.

[196] R.X. Wang, X.M. Tao, Y. Wang, G.F. Wang, Electrical properties and fatigue resistance of polyamide 6,6 fabrics with nanocrystal silver coating, *Journal of Nanoscience and Nanotechnology* 9 (2009) 3062–3066.

[197] Weber, J., Futterer, C., Gowri, V.S., Attia, R., Viovy, J.L., Towards a microfluidic chip for unknown mutation detection and genotyping *La Houille Blanche* 5:40-44 (2006).

[198] Wenz G, *Angew. Chem., Int. Ed. Engl.* 33 (1994) 803.

[199] White LA, Bertoniere NR. Preparation of Cotton/Clay Nanocomposites. Polym 2002;43(2):1279.

[200] White S. B., Jau-Min Shih A. and Pipe K.P., 2011, "Investigation of the electrical conductivity of propylene glycol-based ZnO nanofluids", *Nanoscale Research Letters*, 6, pp: 346.

[201] Wong, Y. W. H.; Yuen, C. W. M.; Leung M. Y. S.; Ku M. Y. S.; Lam H. L. I. (2006). Selected applications of nanotechnology in textiles. *Autex Research Journal 6,* 1- 10 model de articol din revista.

[202] Woodward IS, Schofield WCE, Roucoules V, et al. a Super-hydrophobic Surfaces Produced by Plasma Fluorination of Polybutadiene Films. *Plasma Chem Plasma* P 2006; 26: 507.

[203] Wu J.; Xia, J.; Jing, C.; Lei, W.; Wang, B.P. Formation of hierarchical ZnO nanostructure on tinfoil substrate and the application on wetting repellency, *Applied Physics* A 105 (2011) 221–224.

[204] Yadav A, Virendra P, Kathe AA, Raj S, Yadav D, Sundaramoorthy C, Vigneshwaran N. Functional finishing in cotton fabrics using zinc oxide nanoparticles. *Bull. Mater. Sci.* 2006; 29 (6): 641–645.

[205] Yadav, Richa Srivastava and Alok Kumar.Characterization of ZnO Nanomaterial Synthesized by di_erent methods. *International Journal of Nanotechnology and Applications*, 2007, volume1

[206] Yamamoto O. Change in antibacterial characteristics with doping amount of ZnO in MgO-ZnO solid solution. *Int Inorg Mat* 2001; 2: 643.

[207] Yamamoto, O., Komatsu, M., Sawai, J., Nakagawa, Z. E., Effect of lattice constant of zinc oxide on antibacterial characteristics *J. Mater. Sci. Mater. Med.* 15:847 (2004).

[208] Yang, H.H. Aramid fibers. Bunsel AR, editor. Fibre reinforcements for composite materials. *Composite Materials Series*. Amsterdam: Elsevier Science Publishers; 1988. p. 249–327.

[209] Ye DY, Farriol X. Preparation and characterization of methylcellulose from Miscanthus Sinensis. *Cellul.* 2005; 12, 507.

[210] Yousif, B.F.; Nirmal, U., Wong, K.J. *T*hree-body abrasion on wear and frictional performance of treated betelnut fibre reinforced epoxy (T-BFRE) composite. *Mater Des* 2010;31:4514–21.

[211] Yu, J.; Yang, J.; Liu, B.; Ma, X. Preparation and characterization of glycerol plasticized-pea starch/ZnO-carboxymethylcellulose sodium nanocomposites. *Biores Technol* 2009; 100: 2832–2841.

[212] Yu, M.; Gu, G; Meng, W.D., et al. Superhydrophobic cotton fabric coating based on a complex layer of silica nanoparticles and perfluorooctylated quaternary ammonium silane coupling agent. *Appl Surf Sci* 2007; 253: 3669.

[213] http://test.ttri.org.tw/neweng/RD/images/pub3.pdf

[214] www.malvern.co.uk;

[215] www.anton-paar.com

[216] http://www.znoxide.org/properties.html Physical Properties of Zinc Oxide - CAS 1314-13-2.

[217] International Zinc Association

[218] http://www.aadet.com/article/nanoparticle, Nanoparticle Data

[219] http://www.news-medical.net/health/Properties-of-Nanoparticles.aspx Properties of Nanoparticles

[220] http://www.google.ro/search?q=biogenic+calcites&hl=ro&biw=1366&bih=624&prmd=ivnsb&source=lnms&tbm=isch&ei=fWoHTpqFKNGaOv6S7bgN&sa=X&oi=mode_link&ct=mode&cd=2&ved=0CA0Q_AUoAQ

INDEX

C

D

E

F

G

H

N

O

P

Q

R

S

T

U